# THE WORLD'S H🔥T SPOTS

# Iran

Mikko Canini, *Book Editor*

Bruce Glassman, *Vice President*
Bonnie Szumski, *Publisher*
Helen Cothran, *Managing Editor*
Scott Barbour, *Series Editor*

**GREENHAVEN PRESS**
*An imprint of Thomson Gale, a part of The Thomson Corporation*

THOMSON
GALE

Detroit • New York • San Francisco • San Diego • New Haven, Conn.
Waterville, Maine • London • Munich

*For more information, contact*
Greenhaven Press
27500 Drake Rd.
Farmington Hills, MI 48331-3535
Or you can visit our Internet site at http://www.gale.com

**LIBRARY OF CONGRESS CATALOGING-IN-PUBLICATION DATA**

Iran / Mikko Canini, book editor.
   p. cm. — (The world's hot spots)
   Includes bibliographical references (p. ) and index.
   ISBN 0-7377-1723-8 (lib. : alk. paper)
     1. Iran—History—Revolution, 1979. 2. Iran—Politics and government—
1979–1997. 3. Iran—Politics and government—1997– . I. Canini, Mikko. II. Series.
DS318.8.I654 2005
955.05'4—dc22
                                         2004042500

Printed in the United States of America

# 🔥 CONTENTS

ian organizations Iran supports practice a legitimate
form of resistance to foreign military aggression and
should not be labeled as terrorists.

# 🔥 FOREWORD

The American Heritage Dictionary defines the term *hot spot* as "an area in which there is dangerous unrest or hostile action." Though it is probably true that almost any conceivable "area" contains potentially "dangerous unrest or hostile action," there are certain countries in the world especially susceptible to conflict that threatens the lives of noncombatants on a regular basis. After the events of September 11, 2001, the consequences of this particular kind of conflict and the importance of the countries, regions, or groups that produce it are even more relevant for all concerned public policy makers, citizens, and students. Perhaps now more than ever, the violence and instability that engulfs the world's hot spots truly has a global reach and demands the attention of the entire international community.

The scope of problems caused by regional conflicts is evident in the extent to which international policy makers have begun to assert themselves in efforts to reduce the tension and violence that threatens innocent lives around the globe. The U.S. Congress, for example, recently addressed the issue of economic stability in Pakistan by considering a trading bill to encourage growth in the Pakistani textile industry. The efforts of some congresspeople to improve the economic conditions in Pakistan through trade with the United States was more than an effort to address a potential economic cause of the instability engulfing Pakistani society. It was also an acknowledgment that domestic issues in Pakistan are connected to domestic political issues in the United States. Without a concerted effort by policy makers in the United States, or any other country for that matter, it is quite possible that the violence and instability that shatters the lives of Pakistanis will not only continue, but will also worsen and threaten the stability and prosperity of other regions.

Recent international efforts to reach a peaceful settlement of the Israeli-Palestinian conflict also demonstrate how peace and stability in the Middle East is not just a regional issue. The toll on Palestinian and Israeli lives is easy to see through the suicide bombings and rocket attacks in Israeli cities and in the occupied territories of the West Bank and Gaza. What is, perhaps, not as evident is the extent to which this conflict involves the rest of the world. Saudi Arabia and Iran, for instance, have long been at odds and have attempted to gain control of the conflict by supporting competing organizations dedicated to a

Palestinian state. These groups have often used Saudi and Iranian financial and political support to carry out violent attacks against Israeli civilians and military installations. Of course, the issue goes far beyond a struggle between two regional powers to gain control of the region's most visible issue. Many analysts and leaders have also argued that the West's military and political support of Israel is one of the leading factors that motivated al Qaeda's September 11 attacks on New York and Washington, D.C. In many ways, this regional conflict is an international affair that will require international solutions.

The World's Hot Spots series is intended to meet the demand for information and discussion among young adults and students who would like to better understand the areas embroiled in conflicts that contribute to catastrophic events like those of September 11. Each volume of The World's Hot Spots is an anthology of primary and secondary documents that provides historical background to the conflict, or conflicts, under examination. The books also provide students with a wide range of opinions from world leaders, activists, and professional writers concerning the root causes and potential solutions to the problems facing the countries covered in this series. In addition, extensive research tools such as an annotated table of contents, bibliography, and glossaries of terms and important figures provide readers a foundation from which they can build their knowledge of some of the world's most pressing issues. The information and opinions presented in The World's Hot Spots series will give students some of the tools they will need to become active participants in the ongoing dialogue concerning the globe's most volatile regions.

# ♦ INTRODUCTION

W hen the Boeing 747 bearing Ayatollah Sayyid Ruhollah Musavi Khomeini touched down in Tehran on the morning of February 1, 1979, the future of Iran was deeply uncertain. Mass protests the previous month had ousted Shah (King) Mohammad Reza Pahlavi, whose repressive regime had lasted for more than thirty-five years, and various groups were attempting to fill the vacuum. Khomeini was returning after living fifteen years in Iraq and France, having been exiled by the shah for his antimonarchic views. Despite his absence from Iran, Khomeini's revolutionary message had found its way to Iranians through smuggled tape cassettes, and the 3 million people who turned out that day to witness Khomeini's return were eager to support his vision of a new political order based on Islamic principles. Within weeks, Khomeini was able to seize power. Moving quickly to establish a new government, draft a constitution, and hold a referendum on the creation of the new state, Khomeini announced the establishment of the Islamic Republic of Iran on April 1, 1979.

The Islamic Revolution took the world by surprise. As late as 1977 the U.S. State Department had reported that "the Shah rules Iran free from domestic threat . . . [and] has an excellent chance to rule for a dozen or more years."[1] This misperception cost the United States dearly because in 1979 the shah's pro-West monarchy was transformed into Khomeini's anti-America theocracy. Overnight, Iran, a valuable ally, become a formidable enemy.

Despite the passing of twenty-five years, the relationship between Iran and the United States remains largely antagonistic. The government in Tehran, Iran's capital, refers to America as the "Great Satan" and demands a resumption of economic ties and the release of Iranian funds held by the United States since the revolution before it will consider the possibility of reconciliation. Washington considers Iran to be part of an "axis of evil" (a group including North Korea and Iraq that threatens world peace) and insists on government-to-government dialogue before it will consider lifting its ban on trade and investment. This impasse has been sustained by a combination of mixed political signals, bellicose public rhetoric, and occasional secret dealings and is motivated by both ideological conviction and pragmatic consideration. Furthermore, recent statements by both governments suggest that this stalemate will continue despite the inevitability of further contact as each country maneuvers to define the future of the region.

8

# Foreign Incursions

America's relationship with Iran began in earnest in the years following World War II. At the time, Washington was increasing its participation in world affairs and required dependable oil reserves to supply its growing needs. To this end Iran was invaluable, having both massive reserves and a monarch eager to trade with the West. However, for the Iranian population the presence of American oil companies represented only the most recent incursion of a foreign nation eager to exploit Iranian oil.

The conflict over oil in Iran began in 1901 with the purchase of a concession that guaranteed exclusive rights to explore and drill for oil within Iran by the Anglo-Iranian Oil Company (AIOC), which was controlled by the British government. The terms of the concession were so unbalanced that when the first oil drilling project began in 1908, neither the Iranian government nor the people benefited from their resource. Iranians soon grew to resent the extravagantly wealthy lifestyles of the foreign businessmen and engineers, which contrasted sharply with the poverty of the nation.

The dominance of foreign powers in Iran faced a setback in 1921 when the shah's father, Reza Khan, headed a successful military coup and established himself as the nation's ruler. Reza Khan was eager to end foreign interference in Iran's affairs. He renegotiated agreements and treaties with foreign nations, abolished the special privileges accorded to Europeans in Iran, and took control of the country's finances and communications, which had been largely controlled by foreign powers. These efforts at independence upset those governments with stakes in the country. In August 1941 British and Soviet forces invaded Iran and deposed Reza Khan in favor of his son, who agreed to participate in the new government.

Reza Khan's repressive regime had been deeply unpopular among Iranians, and its collapse led to an increasingly open political system with the new shah. However, foreign intervention remained a sensitive issue because the population still faced economic hardship despite the wealth generated by oil production. This frustration led to the 1951 appointment of Prime Minister Mohammad Mosaddeq, who sought to nationalize the country's oil industry and renegotiate contracts with foreign firms. Using his massive popularity as leverage, Mosaddeq was able to wrest power from the U.S.-backed shah and take control of the country. However, his term as prime minister would last only two years. In August 1953 Mosaddeq's government was toppled by a coup engineered by the American Central Intelligence Agency (CIA), which reinstated the shah.

# The Cleric and the Shah

The close relationship between Tehran and Washington would become one of the main points of contention for those who opposed the shah and perceived the Americans as complicit in his dictatorial rule. This contempt for U.S. involvement in the rule of Iran would become fiercer in 1961, when the shah introduced his "White Revolution," a package of social and economic reforms that sought to modernize the country. The policies mimicked those of his father, who had aggressively attempted to Westernize the nation following World War I. The main result of this earlier enforced modernization had been the creation of a wealthy Westernized class cut off from the general population, which maintained a more traditional Islamic cultural life. Ironically, this largely rural population would begin to migrate into the cities in the 1960s as a direct result of the shah's reforms, ultimately providing support to the clerics who were becoming vocal in their opposition to the government.

By this time, a leading cleric named Ayatollah Khomeini had begun to attract attention for his harsh criticism of the shah, and his arrest in 1962 elevated him to the status of national hero. He second arrest in 1963 led to three days of rioting, which was suppressed by the shah's forces only after more than six hundred demonstrators had been killed. Fearing Khomeini's mass appeal, the shah finally had him exiled in 1964 for his outspoken opposition to a bill that gave U.S. military personnel diplomatic immunity for crimes committed in Iran. Despite improving economic conditions throughout the 1960s and 1970s, animosity toward the shah and his relationship with the United States continued to grow, expressing itself in widespread civil unrest and forcing the shah to adopt increasingly repressive measures in a futile attempt to maintain control. On January 16, 1979, the shah and his family were forced to flee Iran. In less than a month, Khomeini had returned from exile in France and was leading a revolution to end the monarchy.

# Khomeini's Iran

Khomeini's animosity toward foreign, and particularly American, incursions into Iran's political and economic spheres found its outlet in a foreign policy that emphasized a vehement anti-U.S. stance, the elimination of outside influence in the region, the support for Muslim political movements abroad, and increased diplomatic contacts with developing countries. In short, the revolutionary regime sought a reversal of the shah's Western orientation. The transformed foreign policy was to have a profound and immediate effect.

Iran

On November 4, 1979, a group of Iranian students stormed the U.S. embassy in Tehran, spurred on by Khomeini's anti-America speeches. Of approximately ninety people taken hostage, fifty-two remained captive in the embassy for more than a year. This unprecedented event led to the formal end of diplomatic relations between the two countries, the expulsion of most Iranian diplomats from the United States, and the U.S. embargo on Iranian oil. Relations between the two nations became further strained the following April when an American rescue attempt was aborted after eight U.S. Marines were killed in the collision of two U.S. helicopters in the Iranian desert. The animosity fostered during this period of turmoil would resurface with America's involvement in the notorious Iran-Iraq War.

## The Iran-Iraq War

On September 23, 1980, less than a year after the revolution, Iraqi forces attacked Iran in a campaign that was expected to surprise and overwhelm the diminished Iranian military. Iran immediately demanded the withdrawal of Iraqi troops and a return to the borders set by the Algiers Agreement signed by Iran and Iraq in 1975. Iraq refused, and a brutal conflict ensued. Over the next two years Iran was able to repulse the invasion largely thanks to the men and boys who, in a continuation of revolutionary fervor, volunteered to battle the militarily superior Iraqi forces. Iran eventually gained the upper hand, and although given several opportunities to end the conflict, continued to engage Iraq with its sights set on the expansion of the Islamic Republic.

At least ten nations supplied arms to both sides during the conflict, including the United States, who also supplied intelligence and misinformation to both Iran and Iraq. The United States was not primarily motivated by the profits from arms sales but by the desire to contain the two states that each threatened to dominate the region. However, a number of very public events, including Iranian-backed terrorist attacks on U.S. embassies, eventually caused the United States to side with Iraq. As Richard Armitage, the U.S. assistant secretary of defense at the time stated, "while we want no victor, we can't stand to see Iraq defeated."[2] Indeed, the concluding event of the eight-year war would come in July 1988, when an American navy cruiser shot down an Iranian passenger plane in the Persian Gulf, killing 290 people. While the event was described as an accident by American authorities, the Iranians perceived it as a declaration that the United States intended to become more openly involved in the conflict. As historian Dilip Hiro notes, "in the wake of the Iranian airbus disaster, Tehran had two stark choices: either to escalate confrontation with America in the Gulf and/or elsewhere, or to accept unconditionally Security Council Resolution 598 [which called for a cease-fire]. It chose the latter."[3]

# A Call to Arms

After eight years of brutal conflict the war ended without either side having made any real gains. Iran's losses included three hundred thousand dead, a severely damaged economy, and a depleted military. Despite the country's massive financial debt, the next few years saw a dramatic surge in arms purchases as Iran sought to build up its defenses and assert itself as a military power in the region. In addition to traditional arms, Iran also sought chemical, biological, and, it is suspected, nuclear weapons.

Iran began to develop chemical weapons in the first years of the Iran-Iraq War as a response to their repeated use by the Iraqi military. And while there is little information regarding Iran's biological weapons program, the U.S. Defense Department believes that Iran accelerated its development efforts in 1995 after the disclosure of Iraq's program. However, it is Iran's alleged nuclear weapons program that has proven the most contentious. Iran does not currently have any nuclear weapons, and Iranian officials have repeatedly insisted that their burgeoning nuclear program is intended solely for the production of energy.

In a February 1987 address to Iran's Atomic Energy Organization, President Seyed Ali Khamenei (now Iran's "Supreme Leader") stated:

> Regarding atomic energy, we need it now. . . . Our nation has always been threatened from outside. The least we can do to face this danger is to let our enemies know that we can defend ourselves. Therefore, every step you take here is in defence of your country and your evolution. With this in mind, you should work hard and at great speed.[4]

Comments such as these have alarmed U.S. officials who have pressed the International Atomic Energy Agency (IAEA) to demand Iranian compliance with additional protocols, such as random inspections of its nuclear sites, in an attempt to limit the proliferation of atomic weapons in the region. In February 2004 it was discovered that Iran had blueprints for an advanced centrifuge for uranium enrichment that it had withheld from nuclear inspectors.

Iran argues that the United States exercises a double standard concerning nuclear development because it does not object to the existence of Israel's nuclear program and Israel's nuclear research facility is not subject to any inspection or regulation. Indeed, increased conflict between Iran and Israel over the nuclear issue seems likely as Israel has been vocal in condemning the suspected Iranian program and has demonstrated a willingness to intervene and eliminate perceived threats: In 1981 Israeli Defense Force warplanes destroyed a nuclear reactor in Iraq, setting the country's nuclear program behind by a decade.

Hostilities between Iran and Israel can be traced directly back to the 1979 revolution. One of the very first acts of the revolutionary government was to denounce the relationship between the two states that had been fostered under the shah, immediately banning all trade with Israel, especially the sale of oil. Iran's leaders contended that Israel's existence was illegitimate and advocated the eradication of the nation in favor of a reconstituted Palestine. Despite statements such as President Hojjatoleslam Seyed Mohammad Khatami's 1998 avowal that "We have declared our opposition to the Middle East process . . . [but] we do not intend to impose our views on others or to stand in their way,"[5] Iranian animosity toward Israel is still evident today. During a 2003 military parade in Tehran, Iran's new long-range Shehab-3 missiles bore the slogan "Israel must be wiped off the map." However, what concerns most observers is not so much the very public anti-Israel rhetoric, but the less visible state support for terrorist groups such as Hizballah, Hamas, Islamic Jihad, and the Popular Front for the

Liberation of Palestine, all violently opposed to the Arab-Israeli peace process.

# A Twenty-Five Year Stalemate

The threat of Iran's sponsorship of international terrorism and possible development of weapons of mass destruction has shaped America's stance toward the nation. While the George H.W. Bush administration (1989–1993) pursued a policy of "constructive engagement" and lifted some economic sanctions as a reconciliatory gesture, the Clinton administration (1993–2001) took a more hard-line position, referring to Iran as an international outlaw and enforcing more stringent sanctions. Washington's perception of Iran shifted again in 1997 with the election of President Khatami, who was seen as representing a liberalizing force.

The first signs of change came on January 7, 1998, when, in an interview with America's CNN, Khatami called for a "dialogue with the American people," a statement that openly contradicted the regime's previous anti-U.S. policy. This diplomatic gesture was returned six months later when, in a speech announcing the end of a ban on the import of Iranian luxury goods, U.S. secretary of state Madeleine Albright acknowledged the role the United States had played in the 1953 coup. In 2001 Khatami was vocal in condemning the September 11 terror attacks and publicly endorsed the U.S.-led fight against al Qaeda and the Taliban regime in Afghanistan. These gestures have resulted in moments of cooperation between Washington and Tehran, notably in Iran's aid to American forces trying to control radical Islamic groups in Iraq following America's invasion in 2003 and attempts to stem the flow of narcotics from postwar Afghanistan. Further signs of rapprochement came later that same year when the United States offered aid and a temporary easing of sanctions following an earthquake that destroyed Iran's historic city of Bam.

America is also responsible for toppling two of Iran's most problematic neighbors: the Taliban regime in Afghanistan and Saddam Hussein's regime in Iraq. However, these recent military campaigns have also dramatically increased America's presence in the region. Iran now faces American troops on both its eastern and western borders. Additionally, America's various relationships with Israel, Saudi Arabia, Turkey, Pakistan, and the Gulf Emirates indicates that its influence in the Middle East is growing. From Tehran's perspective, the American presence in the region constitutes a threat to its sovereignty, and it accuses the United States of attempting to establish a dominant role in the Middle East through military adventurism and cultural invasion. Washington counters that Iran is working covertly to derail efforts to install democratic governments in Afghanistan and Iraq and has given refuge to fleeing al Qaeda fighters—charges that Tehran denies. Sadly, aggressive posturing and hostile rhetoric accompany every reconciliatory gesture and pacifying statement between Washington and Tehran.

# Inside Iran

Any assessment of U.S.-Iran relations must take into account not only the past half century of acrimony, but also the divided nature of Iranian politics. Iran's governing structure ensures that democratically elected politicians must share power with the unelected hard-line clerical establishment. To complicate matters, a perennial struggle between conservatives committed to the Islamic Republic's founding constitution and reformists eager to introduce democratic change results in the government regularly issuing contradictory statements as each group vies to represent their vision of Islamic Iran.

President Khatami's surprise landslide election in 1997 and reelection in 2001 suggests that the reform movement has finally achieved the momentum necessary to transform not only the nation's domestic politics but also its fraught foreign relations. The reformist foreign policy focuses on the expansion of trade, cooperative security measures, and diplomatic dialogue as a means of advancing Iran's interests and projecting its influence. In short, it represents a move away from Khomeini's ideological foreign policy toward a more pragmatic one. However, while it is true that during Khatami's term Iran has improved its relations with the European Union, Russia, and many Middle Eastern nations, there has been no change in the regime's hostility to the United States. However, the Iranian electorate does not share this enmity. Two-thirds of the Iranian population is younger than thirty and has no memory of life under the shah, no experience of the struggle that led to the Islamic Revolution, and no bitterness about fighting in the Iran-Iraq War. Indeed, a 2002 poll suggests that more than 75 percent of the population would welcome increased ties with America.

Despite this popular support, the inability of reformists to produce substantial change in the six years since Khatami's first election has led to the perception that the reform movement has failed. Political activists are still being jailed and proreform newspapers are being shut down. This sense of failure has created widespread apathy, evidenced by the low voter turnout in the February 2004 parliamentary elections, which allowed conservatives to regain control of the parliament. The clerics are unlikely to permit a credible reformist candidate stand in the 2005 presidential election, and the future of the reform movement in Iran is in doubt.

While some have suggested that Iran's conservatives have systematically subverted the efforts of the reformists to improve relations with America, there are signs that more pragmatic conservatives are also interested in developing a functioning relationship with Washington. This shift within the conservative camp has led analysts to speculate that the regime is slowly adopting the "China model," whereby opposition to political repression is softened by relaxed social constraints, economic development, and increased foreign investment. A more pragmatic regime and improved relations with America could well lead to a more stable and prosperous Iran; however, it may not be the democratic Iran that the reforms have struggled to create.

# Notes

1. Quoted in James A. Bill, *U.S.-Iran Relations: Forty Years of Observations*, Middle East Institute, February 20, 2004. www.mideasti.org.
2. Quoted in Dilip Hiro, *The Longest War: The Iran-Iraq Military Conflict*. New York: Routledge, 1991, p. 186.
3. Hiro, *The Longest War*, p. 240.
4. Quoted in Vitaly Fedchenko et al., *Iran's Nuclear Program and Russian-Iranian Relations*, Institute for Applied International Research, February 2003. www.iair.ru.
5. Interview with Iranian president Mohammad Khatami, CNN, January 7, 1998. www.cnn.com.

# CHAPTER 1

# From Revolution to Reform

# The Islamic Revolution

## By Shaul Bakhash

*By 1979 Iran was in the midst of a dramatic transformation as the British-backed ruler of Iran, Reza Shah, modernized the country, introducing a telephone system, a railroad, modern schools, and a university. However, rapid economic and social modernization came at a cost. Increased inflation and a widening income gap meant that although foreign businesses continued to profit from Iran's rich natural resources, most Iranians did not benefit economically. In addition, the shah introduced policies, such as modern legal codes and the banning of the veil for women, that were meant to undermine Islamic traditions. The Iranian revolution of 1979 was the result of mass discontent with the dictatorial rule of Reza Shah, the foreign control of Iran's oil, and the erosion of traditional Islamic values. The charismatic ayatollah Ruhollah Khomeini, with his vision of the world's first theocratic state, united the population in a successful revolt against the government. In the following article, Shaul Bakhash, a professor of Middle East history at George Mason University, argues that the shah's erratic responses to Iranian demands for change forced the population to become increasingly radical and ultimately led to his overthrow. Bakhash serves on the Middle East advisory board of Human Rights Watch and on the editorial boards of the* Journal of Democracy *and the* Middle East Journal. *He is the author of* Iran: Monarchy, Bureaucracy, and Reform Under the Qajars, 1858–1896, *and* The Politics of Oil and Revolution in Iran.

The Iranian Revolution in 1979 astonished the world because an opposition armed only with slogans and leaflets overthrew a ruler with formidable assets at his disposal. The Shah [Mohammad Reza Pahlavi] commanded an army of 400,000, a large police force, and a fearsome secret police, Savak, with 4,000 full-time agents and scores

Shaul Bakhash, *The Reign of the Ayatollahs: Iran and the Islamic Revolution*. London: I.B. Tauris & Co., Ltd., 1985. Copyright © 1985 by BasicBooks, Inc. All rights reserved. Reproduced by permission.

of part-time informers. The government controlled the mass media and kept a tight rein on the press. There was only one officially sanctioned political party, and it was subservient to the monarch.

Moreover, the revolution took place against the background of nearly two decades of impressive economic growth. Using a steady flow of oil revenues, the Shah had built roads, dams, railroads, and ports; he had established steel and petrochemical industries; he had helped an entrepreneurial private sector develop a range of consumer industries.

A slight downturn in oil income in 1977 did not alter the fact that the country had strong foreign exchange reserves, investments abroad, few foreign debts, and an inflow of oil revenues that, compared to the Iranian situation only four years earlier, still constituted a cornucopia of plenty. In addition, Iran's borders were secure and it dominated the region. Until nearly the very end, the Shah had the support of both great powers [the United States and the Soviet Union], the western European states, his immediate neighbors, and the Arab states of the Persian Gulf.

# The Roots of Revolution

Recent studies have emphasized the complex nature of the background to the Iranian revolution. The Pahlavi dynasty, established in 1925, did not have deep roots in the country, and the Shah almost lost his throne in 1953. At that time, he was restored to power by an army coup, engineered with the assistance of the CIA. In 1963 widespread riots again shook the country. They were inspired by Ruhollah Khomaini, a religious leader who was just rising to preeminence, and they were put down with great severity. The Shah gained support in the 1960s, thanks to a program of land distribution, a number of other reforms, and a decade of sustained economic growth. By the eve of the revolution, however, this credit had been dissipated by the Shah's autocratic tendencies, the dislocations caused by a reckless economic program, and policies that alienated important sectors of the community.

But between 1963 and 1977, even as he carried out reforms, the Shah steadily reinforced the foundations of a royal autocracy. He suppressed the independent political parties and founded court-sponsored political movements. He packed Parliament with yes-men and he muzzled the press. He extended government control over such organizations as labor unions and trade guilds. He avoided the older, independently minded officials of his youth—men like Ali Amini, Abol Hasan Ebtehaj, and Abdollah Entezam—and surrounded himself either with sycophants or with technocrats who were competent but who lacked an independent political base of their own.

One result of these developments was to push elements of the opposition toward an increasingly radical position. The suppression of the 1963 protest movement persuaded young men of the National Front that constitutional methods of opposition against the Shah were ineffective. The National Front (NF), a coalition of parties headed by the nationalist prime minister, Mohammad Mossadegh, during the oil nationalization movement of 1950–53, was in 1963 still the major opposition political party. Two groups broke away to form what later became the Mojahedin-e Khalq and the Fadayan-e Khalq guerrilla movements, both dedicated to the violent overthrow of the regime. Secret movements emerged among the seminary students and younger clerics of Qom, a major religious center, also dedicated to overthrowing the regime.

The propensity of Savak to extend its operations, the desire of the Shah and his bureaucracy to impose state control over universities, private schools, business groups, religious endowments, and numerous other private organizations, meant that citizens who would not normally concern themselves with politics found the bureaucracy increasingly involved in their lives. In 1975, the Shah abolished political parties, including the Iran Novin which he had himself sponsored, and announced the establishment of a single party, Rastakhiz (Resurgence), for all Iranians. Those who did not wish to be part of the political order, he remarked, could take their passports and leave the country. Civil servants, university professors, and ordinary citizens were pressed to join the party. For the first time, nonpolitical individuals were being required to declare themselves and publicly identify with a royal political party.

Moreover, the Shah's rule appeared increasingly arbitrary. Although Parliament was a rubber stamp, he preferred to rule by imperial decree, in direct contravention of the constitution. He personally announced the nationalization of the secondary schools and he ordered industrialists to sell 49 percent of their shares to their workers. In both cases, implementation began long before the proper legislation was approved by Parliament. The leading officers of the government took their cue from the ruler. In order to make way for new streets and avenues, the mayor of Tehran sent bulldozers to demolish private homes and working-class districts, and in 1977 riots occurred as a result of this policy. To create agroindustries, the government forced villagers to sell their farmland and often razed entire villages. The minister of agriculture, intent on creating agricultural "poles" where rural populations could be concentrated and agricultural services more efficiently delivered, secured approval from Parliament for a law which permitted him to transfer the population of whole villages from one district to another.

# Protests Begin

A dimension of lawlessness was added to this arbitrary system of rule in 1977, when the first public protests in many years began to be voiced against the regime. A woman university professor who had participated at politically charged poetry-reading sessions in November 1977 was taken to an empty lot by agents of the police and beaten. Bombs, too small to cause great damage, but large enough to serve as a warning, were placed in the homes and offices of several of the lawyers active in the Iranian Committee for the Defense of Freedom and Human Rights. When members of opposition groups met for discussions in a private garden outside Tehran, "workers" showed up and beat up the participants, many of whom were middle-aged. One analyst of the Iranian revolution has noted that the Shah generated in Iranians a sense of humiliation and ultimately rage; ". . . the behavior of the Shah increasingly came to be experienced as an insult—a narcissistic injury to his own people. . . . He showed the Iranians no compassion and no empathy."

The period between 1963 and 1973 was a decade of rapid economic growth, which led to expanding job and education opportunities, improved standards of living, and rising patterns of consumption. More Iranians were able to buy radios, cars, refrigerators, or shoes, but the benefits of the economic boom were not evenly spread. People in some social strata did far better than those in other strata; the urban centers benefited more than the countryside. Government credit policy tended to favor the large industrialist and farmer as against the workshop owner or small cultivator. Housing conditions in large urban centers declined for the lower income groups. However, as long as the majority saw a chance to improve their condition at a reasonable pace, these problems did not seem likely to generate insuperable political problems. It was the explosion in oil prices in 1974 that severely dislocated both economic and social life.

# Frustrated Aspirations

Virtually overnight, Iran's oil revenues quadrupled, from under $5 billion to nearly $20 billion a year. The Shah believed that this money would enable him at last to carry the country to his long promised "Great Civilization" and within a decade, turn Iran into one of the world's five leading industrial countries. He plunged into a reckless spending program.

The results were predictable: the economy overheated, prices of housing, food, and basic necessities soared. Rural migrants drained the countryside of agricultural labor and swelled the shantytowns and the urban underclass of the large cities. Bottlenecks developed in all sectors of the economy. Ships waited months for their turn to unload cargo at Iranian ports; there were shortages of cement and steel for home construction. The shortage of skilled labor necessitated the bringing in of tens of thousands of foreign workers. There were massive electricity shortages. Iranians remarked that the power failures and blackouts marked the arrival of the Great Civilization.

Moreover, if real incomes for workers and the white-collar salaried employees steadily improved in the 1963–73 period, after 1974 the economic position of these groups deteriorated. Resentment was intensified by the widening gap in incomes, by the ability of the privileged few to make fortunes by dealing in land, scarce commodities, and goods, and through commissions on large and questionable government contracts. . . .

The government decided to deflate the economy by cutting back on investment, curtailing projects, and stopping new hiring in the civil service. The sudden about-face, after a period of uncontrolled spending, led to a business downturn and reduced employment and business opportunities. It also burst the balloon of inflated expirations. The purchase of a house, or even a car, was suddenly beyond the reach of middle-class families; bazaar merchants were faced with dwindling business opportunities; clerical staff on fixed salaries felt bitterly resentful and betrayed. The pressure of frustrated aspirations triggered the political crisis that followed.

## Protests Increase

The economic crisis coincided with pressure on the Shah from such organizations as Amnesty International and the International Commission of Jurists on the condition of prisons and treatment of political prisoners in Iran. More important was the pressure for human rights reforms emanating from the [U.S.] Carter Administration. The Shah was sensitive to such pressure, particularly from a Democratic administration. In 1961–62, during another period of internal crisis, he had been persuaded by the Kennedy Administration to appoint a reformist prime minister and to implement land reform. It was an experience the Shah did not soon forget. In 1977, he saw the Carter human rights campaign as a repeat performance of his experience with Kennedy. This time, he took the initial measures on his own initiative. For example, he introduced new regulations that permitted civilian defendants brought before military tribunals to be represented by civilian lawyers and to enjoy open trials. The Shah also slightly eased press controls.

Members of the intelligentsia, professional groups, and leaders of the middle-class opposition parties were quick to take advantage of this slight opening. In May 1977, fifty-three lawyers addressed a letter to the imperial court, demanding an independent judiciary. In June, three leaders of the National Front wrote directly to the Shah asking for a restoration of press freedoms, the implementation of the constitution, and the freeing of political prisoners. A group of forty writers and intellectuals wrote to the prime minister, Amir-Abbas Hoveyda, to protest censorship and the suppression of intellectual freedom. Ninety-eight intellectuals signed a second letter to the prime minister in the same vein. . . .

Protests entered a new phase in January 1978, when seminary students in the holy city of Qom took to the streets to object to a government-inspired article in the newspaper, *Ettelaat*, that cast aspersions on the character of Ayatollah Ruhollah Khomaini. Khomaini, who had a large following among the seminarians, had been expelled from Iran in 1964 for his attacks on the Shah and was living in exile in Iraq. A confrontation with the authorities during the Qom demonstrations led to a number of deaths.

The Qom clashes sparked a series of mourning ceremonies, processions, and riots that over the next twelve months shook dozens of towns and cities. In February 1978, mourning ceremonies for the Qom dead were observed in half-a-dozen cities. But in Tabriz, a young man was shot and there were severe riots. In March, mosque services and processions were organized in fifty-five urban centers; in half-a-dozen towns the ceremonies turned violent. In May, there were more demonstrations and riots.

# "Black Friday"

This new phase of the opposition movement differed significantly from what had come before. The earlier protests were led by the intelligentsia and the middle classes, took the form of written declarations, and were organized around professional groups and universities. The new protests were led by clerics, were organized around mosques and religious events, and drew for support on the urban masses. While the earlier protests were concentrated in Tehran, this new phase spread the protests to the entire country. The earlier protests were generally reformist in content, seeking a redress of grievances and the implementation of the constitution. The mosque-led demonstrations were more radical, even revolutionary in intent. It has been alleged that it was only in the later stages of the protest movement that the Shah was personally denounced, demands made for his overthrow and for the establishment of an Islamic form of government, and Khomaini treated as the leader of the opposition forces. In fact, these themes appear in the pamphleteering literature of the mosque-led protests as early as

February 1978. During the year of protests in 1978, it was adherents of Khomaini, and the proponents of a radical solution, who rapidly gained the upper hand.

The clerics displayed their ability to mobilize the people on 4 September, when they organized mass prayers to mark *id-e fetr* (the end of the Ramadan fasting period). In Tehran, almost 100,000 came together for the communal prayer, then marched to Shahyad Square, shouting pro-Khomaini slogans. For the next three days demonstrations continued, growing larger in size and more radical in their slogans. On 7 September, demonstrators openly called for an Islamic government, denounced the Shah, and repeated the slogan, "Khomaini is our leader." The government, alarmed by the size and the radical temper of the demonstrations, and by indications that the demonstrators were attempting to subvert the troops, declared martial law on the night of 7–8 September. The next day, at Jaleh Square in the working-class district of Tehran, demonstrators, unaware of the martial law regulations, refused to disperse; troops opened fire, and large numbers of demonstrators were killed.

The Jaleh Square massacre, which became known as "Black Friday" in the folklore of the revolution, was a turning point in the protest movement. Compromise with the Shah became extremely difficult if not impossible after this date, and the moderates found themselves forced to take a more radical, uncompromising stand.

Demonstrations continued after 8 September. In October, the first of the strikes in the public sector occurred, and they spread quickly. By November, workers in the oil industry, the customs department, the post office, government factories, banks, and newspapers were on strike. These strikes crippled and finally paralyzed the economy.

# The Royal Response

The Shah's response to these developments was uncertain and erratic. He alternated between concession and clampdown; but neither the periods of clampdown nor the concessions he made were effective, given the crisis in the country and the demands of the opposition. In response to the initial riots, he dismissed unpopular officials, released political prisoners, announced free elections, and promised a Western-style democracy.

In August 1978, following a fire at the Rex Cinema at Abadan in which 477 people lost their lives, he removed Jamshid Amuzegar and appointed Ja'far Sharif-Emami as prime minister. Sharif-Emami made several concessions to the clerics. He set aside the "monarchic" calendar, imposed two years earlier and based on the date Cyrus founded the Persian empire; he closed down gambling casinos and nightclubs, abolished the post of minister of state for women's affairs, set up a ministry for religious affairs, released jailed clerics from prison, and permitted the great prayer meeting on *id-e fetr*. He made concessions to the secular opposition by lifting press censorship, permitting freer debate in Parliament, and allowing renewed activity by political parties.

However, Sharif-Emami proved unable to restore order or to quell demands for more radical change, and in November 1978 the Shah replaced him with a military man, General Gholam-Reza Azhari. The new prime minister, in turn, announced tougher measures against rioters, strikers, and violators of marital law regulations. On the other hand, the Shah went on national television, referred to himself as *padishah* (king) rather than the title he always demanded, *shahanshah* (king of kings), told the people he had heard their "revolutionary message," and promised to correct past mistakes. Presumably to placate public opinion, he allowed the arrest of 132 government leaders, including the former prime minister, Amir Abbas Hoveyda, and the former Savak chief, General Ne'matollah Nasiri.

The Shah's uncertain response to the crisis is attributable to a number of factors, including the fact that he had never been decisive under pressure, and the situation was inherently difficult. By September 1978, it was unclear whether even a massive crackdown could end the protests; and in any case the Shah was reluctant to use greater force and cause more bloodshed. Moreover, he was receiving conflicting advice both from his own advisers and from Washington.

Throughout the crisis, he waited for the United States to tell him what to do. But in Washington counsels were divided: the State Department, under Cyrus Vance, believed the Shah should negotiate with the opposition. The National Security Adviser, Zbigniew Brzezinski, believed that the Shah should be told he would have U.S. support for whatever measures he thought necessary to restore order. The message the Shah received, through his own ambassador to Washington, Ardeshir Zahedi, and through the American ambassador in Tehran, William Sullivan, was thus conflicting, and this added to the ruler's paralysis and indecision.

# The Shah Is Exiled

In December, the Shah finally decided to deal with the opposition. He invited the National Front leader, Karim Sanjabi, to the palace with a view to offering him the government. But Sanjabi was already bound by an agreement with Khomaini which did not recognize the Pahlavi monarch as the legitimate ruler of the country. The National Front leader wanted the Shah to leave the country, but the Shah refused.

By the end of December, however, the Shah's position had become untenable, and the British and American ambassadors were urging him to go abroad. The Shah now turned to another member of the National Front, Shapour Bakhtiar. Bakhtiar was committed to a constitutional transfer of power; he had little use for clerical rule. He agreed to accept the prime ministership from the Shah and to remain loyal to the constitution, on condition the Shah handed over authority to a Regency Council and left the country on a "vacation" of undetermined length. The Shah left Iran on 16 January 1979.

Bakhtiar acted with energy. He dissolved Savak, gave freedom to the press, and announced he would sever diplomatic relations with Israel and South Africa. He sought desperately to maintain calm on the streets. Fearing an army coup, he begged Khomaini not to return to Iran yet (even closing the airport to prevent Khomaini's return), and offered to go to Paris himself to talk to the Imam [Khomaini].

However, Bakhtiar lacked power on the streets. Three days after the Shah's departure, a million people marched in Tehran demanding Bakhtiar's resignation. At Khomaini's instruction, employees in ministries refused to let Bakhtiar's ministers into their offices. When he arrived back in Tehran on 1 February 1979, Khomaini appointed his own prime minister, Mehdi Bazargan. It was only a matter of time before the government of Bakhtiar would collapse, and the end came on 11 February. The revolutionary forces took control, and Khomaini triumphantly announced the establishment of the Islamic state.

# A Call for Islamic Government

## By Ayatollah Khomeini

*Ayatollah Ruhollah Khomeini became well known in the early 1960s for his opposition to the U.S.-backed regime of Reza Khan (later known as Reza Shah). Khomeini's outspoken attitude eventually led to his arrest and deportation in 1964. He lived in Iraq for the next fifteen years, where he continued to encourage mass demonstrations in Iran with calls for the creation of an Islamic state. After Reza Shah fled Iran in 1979, Khomeini returned to the country and was declared political and religious leader of the newly created Islamic Republic. He reinstated Islamic law and suppressed his opposition. He continued to serve as de facto leader of Iran until his death in 1989. After his death the Islamic state became increasingly moderate. In the following selection, Khomeini offers two arguments for the establishment of an Islamic government. The first is that the foundational laws of Islam, which provide the model for an ideal society, require enforcement by a strong Islamic government. He points to the prophet Muhammad's establishment of an Islamic state as an important precedent. Khomeini's second argument is that in order to protect themselves against Western colonialism, Muslims need to unite under a common leadership to safeguard not only Islamic values but also their economic prosperity.*

A body of laws alone is not sufficient for a society to be reformed. In order for law to ensure the reform and happiness of man, there must be an executive power and an executor. For this reason, God Almighty, in addition to revealing a body of law (i.e., the ordinances of the *shari'a*), has laid down a particular form of government together with executive and administrative institutions.

The Most Noble Messenger [Muhammad] (peace and blessings be upon him) headed the executive and administrative institutions of Mus-

Ayatollah Khomeini, "The Necessity for Islamic Government," *Islam and Revolution*, by Imam Khomeini, translated by Hamid Algar. London: KPI Limited, 1985. Copyright © 1981 by Mizan Press. Reproduced by permission.

lim society. In addition to conveying the revelation and expounding and interpreting the articles of faith and the ordinances and institutions of Islam, he undertook the implementation of law and the establishment of the ordinances of Islam, thereby bringing into being the Islamic state. He did not content himself with the promulgation of law; rather, he implemented it at the same time, cutting off hands and administering lashings and stonings. After the Most Noble Messenger, his successor had the same duty and function. When the Prophet appointed a successor, it was not for the purpose of expounding articles of faith and law; it was for the implementation of law and the execution of God's ordinances. It was this function—the execution of law and the establishment of Islamic institutions—that made the appointment of a successor such an important matter that the Prophet would have failed to fulfill his mission if he had neglected it. For after the Prophet, the Muslims still needed someone to execute laws and establish the institutions of Islam in society so that they might attain happiness in this world and the hereafter.

By their very nature, in fact, law and social institutions require the existence of an executor. It has always and everywhere been the case that legislation alone has little benefit: legislation by itself cannot assure the well-being of man. After the establishment of legislation, an executive power must come into being, a power that implements the laws and the verdicts given by the courts, thus allowing people to benefit from the laws and the just sentences the courts deliver. Islam has therefore established an executive power in the same way that it has brought laws into being. The person who holds this executive power is known as the *vali amr.*

The Sunna [a text outlining Muslim traditions] and path of the Prophet constitute a proof of the necessity for establishing government. First, he himself established a government, as history testifies. He engaged in the implementation of laws, the establishment of the ordinances of Islam, and the administration of society. He sent out governors to different regions; both sat in judgment, himself and appointed judges; dispatched emissaries to foreign states, tribal chieftains, and kings; concluded treaties and pacts; and took command in battle. In short, he fulfilled all the functions of government. Second, he designated a ruler to succeed him, in accordance with divine command. If God Almighty, through the Prophet, designated a man who was to rule over Muslim society after him, this is in itself an indication that government remains a necessity after the departure of the Prophet from this world. Again, since the Most Noble Messenger promulgated the divine command through his act of appointing a successor, he also implicitly stated the necessity for establishing a government. . . .

# The Foundation of Islamic Government

The nature and character of Islamic law and the divine ordinances of the *shari'a* furnish additional proof of the necessity for establishing government, for they indicate that the laws were laid down for the purpose of creating a state and administering the political, economic, and cultural affairs of society.

First, the laws of the *shari'a* embrace a diverse body of laws and regulations, which amounts to a complete social system. In this system of laws, all the needs of man have been met: his dealings with his neighbors, fellow citizens, and clan, as well as children and relatives; the concerns of private and marital life; regulations concerning war and peace and intercourse with other nations; penal and commercial law; and regulations pertaining to trade and agriculture. Islamic law contains provisions relating to the preliminaries of marriage and the form in which it should be contracted, and others relating to the development of the embryo in the womb and what food the parents should eat at the time of conception. It further stipulates the duties that are incumbent upon them while the infant is being suckled, and specifies how the child should be reared, and how the husband and the wife should relate to each other and to their children. Islam provides laws and instructions for all of these matters, aiming, as it does, to produce integrated and virtuous human beings who are walking embodiments of the law, or to put it differently, the law's voluntary and instinctive executors. It is obvious, then, how much care Islam devotes to government and the political and economic relations of society, with the goal of creating conditions conducive to the production of morally upright and virtuous human beings. . . .

Second, if we examine closely the nature and character of the provisions of the law, we realize that their execution and implementation depend upon the formation of a government, and that it is impossible to fulfill the duty of executing God's commands without there being established properly comprehensive administrative and executive organs. . . .

# Muslims Unite

In order to assure the unity of the Islamic *umma* [community of Muslims], in order to liberate the Islamic homeland from occupation and penetration by the imperialists and their puppet governments, it is imperative that we establish a government. In order to attain the unity and freedom of the Muslim peoples, we must overthrow the oppressive governments installed by the imperialists and bring into existence an Islamic government of justice that will be in the service of the people. The formation of such a government will serve to preserve the disciplined unity of the Muslims; just as Fatimat az-Zahra [Muham-mad's daughter] (upon whom be peace) said in her address: "The Ima-mate [a region under Islamic rule] exists for the sake of preserving or-der among the Muslims and replacing their disunity with unity."

Through the political agents they have placed in power over the people, the imperialists have also imposed on us an unjust economic order, and thereby divided our people into two groups: oppressors and oppressed. Hundreds of millions of Muslims are hungry and deprived of all form of health care and education, while minorities comprised of the wealthy and powerful live a life of indulgence, licentiousness, and corruption. The hungry and deprived have constantly struggled to free themselves from the oppression of their plundering overlords, and their struggle continues to this day. But their way is blocked by the rul-ing minorities and the oppressive governmental structures they head. It is our duty to save the oppressed and deprived. It is our duty to be a helper to the oppressed and an enemy to the oppressor. This is noth-ing other than the duty that the Commander of the Faithful [Muham-mad's cousin Ali b. Abi Talib] (upon whom be peace) entrusted to his two great offspring in his celebrated testament: "Be an enemy to the oppressor and a helper to the oppressed.". . .

How can we stay silent and idle today when we see that a band of traitors and usurpers, the agents of foreign powers, have appropriated the wealth and the fruits of labor of hundreds of millions of Mus-lims—thanks to the support of their masters and through the power of the bayonet—gaining the Muslims not the least right to prosperity? It is the duty of Islamic scholars and all Muslims to put an end to this system of oppression and, for the sake of the well-being of hundreds of millions of human beings, to overthrow these oppressive govern-ments and form an Islamic government. . . .

If the ordinances of Islam are to remain in effect, then, if en-croachment by oppressive ruling classes on the rights of the weak is to be prevented, if ruling minorities are not to be permitted to plunder and corrupt the people for the sake of pleasure and material interest, if the Islamic order is to be preserved and all individuals are to pursue the just path of Islam without any deviation, if innovation and the ap-proval of anti-Islamic laws by sham [foreign controlled-]parliaments are to be prevented, if the influence of foreign powers in the Islamic lands is to be destroyed—government is necessary. None of these aims can be achieved without government and the organs of the state. It is a righteous government, of course, that is needed, one presided over by a ruler who will be a trustworthy and righteous trustee. Those who presently govern us are of no use at all for they are tyrannical, corrupt, and highly incompetent.

In the past we did not act in concert and unanimity in order to establish proper government and overthrow treacherous and corrupt rulers. Some people were apathetic and reluctant even to discuss the theory of Islamic government, and some went so far as to praise oppressive rulers. It is for this reason that we find ourselves in the present state. The influence and sovereignty of Islam in society have declined; the nation of Islam has fallen victim to division and weakness; the laws of Islam have remained in abeyance and been subjected to change and modification; and the imperialists have propagated foreign laws and alien culture among the Muslims through their agents for the sake of their evil purposes, causing people to be infatuated with the West. It was our lack of a leader, a guardian, and our lack of institutions of leadership that made all this possible. We need righteous and proper organs of government; that much is self-evident.

# The Iran-Iraq War

By Gary Sick

*When Iraq attacked Iran on September 22, 1980, the Islamic Republic was barely a year old and was considered a weak target by the militarily superior Iraq. However, thanks largely to the Basij, a volunteer paramilitary organization made up of poorly trained armed boys and older men, Iran was able to repulse the invasion. Rejecting several peace treaties, Iran went on the offensive. However, by 1987 Iraq was receiving support from the United States and its allies, and a defeated Iran was forced to accept a cease-fire in 1988. In the following selection, Gary Sick describes the events that led to the Iran-Iraq War and its consequences for Iran. Iran's action during the war isolated the country from the international community. In addition, the war greatly reduced Iran's revolutionary zeal. Sick served on the National Security Council under presidents Ford, Carter, and Reagan, and was the principal White House aide for Iran during the Iranian Revolution. He is the author of* All Fall Down: America's Tragic Encounter with Iran *and* October Surprise: America's Hostages in Iran and the Election of Ronald Reagan.

On September 22, 1980, the government of Iraq launched simultaneous strikes against all Iranian airfields within reach of its bombers, while its massed armies advanced along a 450-mile front into Iran's Khuzistan Province. On August 20, 1988, one month short of the war's eighth anniversary, the guns fell silent after the government of Iran, with great reluctance and after a full year of equivocation, accepted a United Nations cease-fire proposal—an act which the [de facto leader of Iran] Ayatollah Ruhollah Khomeini characterized as "more deadly than taking poison." After eight years of brutal conflict, the forces of these two bitterly hostile foes ended the fighting very close to where they had begun. Neither side achieved its war aims. . . .

## The 1975 Algiers Agreement

Reduced to its essentials, the Iran-Iraq war was a dispute about borders, specifically a disagreement about the division of the waters of

Gary Sick, "Trial by Error: Reflections on the Iran-Iraq War," *Middle East Journal*, Spring 1989.

the Shatt al-Arab River—the confluence of the Tigris and Euphrates rivers—which separates the two countries in the south. Although control of the river was a matter of contention going back to Ottoman times [1534–1918] and beyond, the modern dividing line between the two independent states was first established in a border treaty in 1937. It provided for Iraq to control the river up to the bank on the Iranian side—with the exception of an eight-kilometer stretch before [the Iranian island] Abadan where the *thalweg* principle (centerline of the navigable channel) would apply. (Reza Shah [Pahlavi, ruler of Iran] reportedly later told his son that he regarded this as an error in judgment and regretted his acceptance.) In 1969, after the Baath Party assumed power in Baghdad, Iran unilaterally renounced the 1937 treaty and systematically began to challenge Iraqi control of the Shatt al-Arab by not flying the Iraqi flag on its ships and by refusing Iraqi pilots.

The issue appeared to be resolved in 1975, when Muhammad Reza Shah Pahlavi met with then Vice President Saddam Hussein of Iraq on the margins of the meeting of the Organization of Petroleum Exporting Countries (OPEC) in Algiers and surprised the world—and even their close friends and advisers—by adopting the thalweg as the dividing line along the entire river boundary. For several years previously, the shah, together with Israel and the United States, had supported a Kurdish insurrection in northern Iraq. In return for acceptance of the thalweg, the shah agreed to terminate support for the Kurds. This he promptly proceeded to do, dealing the *coup de grâce* to the Kurdish resistance. Observance of the treaty provisions by both Iran and Iraq was satisfactory from 1975 until the revolutionary takeover in Iran in February 1979. At that time, however, Iran still occupied several small pockets of land that were to be ceded to Iraq under the treaty.

# Origins of the War: Iran

In the exuberant atmosphere following the overthrow of the shah [in 1979], Iranian leaders displayed no interest in diplomatic niceties, much less support for any legacies of the shah's rule. Although Iran did not formally renounce the 1975 treaty, neither did it offer assurances that the treaty would be observed. On the contrary, revolutionary leaders almost casually let it be known that they did not consider themselves bound by any of the shah's agreements. Instead, they pointedly noted that in traditional Islam there were no borders dividing the faithful. Those remarks, when coupled with fiery rhetoric calling for export of the revolution to all of the Islamic world, gave Iraq and other neighbors of Iran justifiable grounds for concern.

From the beginning, Iran's hostility to the Iraqi regime had, in addition to its general revolutionary and ideological zeal, a personal edge to it. Khomeini had spent 13 years of his exile in the holy city of Najaf in Iraq, and he was well acquainted with the concerns of the Shi'i population and with the secular nature of Baathist rule. Khomeini was unceremoniously ejected from Iraq in October 1978 in response to the shah's complaints about his political activities, and the ayatollah viewed this as evidence of Saddam Hussein's sympathy with his arch enemy. Revolutionary Iran ignored Iraq's early efforts to develop satisfactory relations, and Iranian leaders made no secret of their support for the Shi'i opposition in Iraq.

Despite this, there were no major incidents between the two countries during the first year of the revolution. The situation began to deteriorate sharply in early 1980 when the Iraqi government arrested Ayatollah Muhammad Baqir al-Sadr, an internationally respected Shi'i clergyman. In early April 1980, bombs exploded at several locations in Iraq and there were attempts on the lives of two Iraqi officials, Tariq Aziz and Latif Nasif Jasim (who were later to become foreign minister and information minister respectively). Iraq suspected, probably rightly, that these events were the work of pro-Iranian opposition forces. Shortly thereafter, Ayatollah Baqir al-Sadr and his sister were reported to have been executed in prison by the Iraqis, setting off emotional shock waves in Shi'i circles in Iran and elsewhere.

In retrospect, it is evident that the events of April 1980 represented the crucial turning point that eventually led to war. Curiously, there is no evidence that either party protested interference in its domestic affairs by the other, although such activity was clearly prohibited by the 1975 treaty. Moreover, Iran, which was absorbed in its own revolutionary politics, seems to have ignored the danger signs and made no effort to reconstitute its military forces. On the contrary, as a result of repeated purges, Iran's military was in a state of almost total disarray.

Thus, Iran's behavior in the immediate post-revolutionary period left it with the worst of all possible worlds. Its rhetoric and meddling with the Shi'i opposition in Iraq was highly provocative, while its military weakness made it a tempting target. The combination proved deadly.

# Origins of the War: Iraq

Iraq has claimed that its invasion of Iran on September 22, 1980, was carried out in self-defense, citing the well-known formula of the *Caroline* case in international law—that there was "a necessity of self-defence, instant, overwhelming, leaving no choice of means and no moment of deliberations." The reality appears otherwise.

Iraq claims that Iranian aircraft violated Iraqi air space on 69 occasions between April and September 1980 and that on September 4 Iranian artillery opened fire across the Iraqi border from the three small parcels of land that were supposed to be returned to Iraq under the 1975 treaty. Assuming the accuracy of these charges, the subsequent Iraqi attack, which bombed targets throughout Iran and captured more than 4,000 square miles of Iran's Khuzistan Province, would appear to be disproportionate to the provocation. Iraq never claimed that Iran was massing forces, and the total absence of any Iranian military preparation was unmistakably obvious in the first few weeks of the war.

The available evidence suggests that Iraq conducted a systematic buildup of its military forces between April and September 1980 in preparation for a lightning offensive. On September 7, Iraq sent its first warning note to Iran and simultaneously sent troops to capture one of the disputed parcels of land. Other disputed areas were reclaimed in successive days, so that by September 13 the Iraqi chief of staff could declare that, "we have regained all the land areas which have been trespassed upon by the Iranian side and have settled our dispute with Iran concerning the land differences."

Four days later, in a speech to the nation, President Saddam Hussein announced that, "Since the rulers of Iran have violated this accord . . . I here announce before you that the Accord of March 6, 1975, is terminated on our part too. Therefore, the legal relationship in the Shatt al-Arab must return as it had been prior to March 6, 1975." He also renounced any claim on Iranian territory. In succeeding days, Iraq reported incidents along the river in the vicinity of Basra, and on September 22 its forces attacked.

# Iraq's Goals

Given this sequence of events, Iraq's claims of urgent self-defense are less than totally convincing. Observers in the region and elsewhere interpreted these claims as a diplomatic fig leaf to justify an attempt to overthrow a revolutionary regime in Iran that posed a serious threat to Iraqi internal stability. Iraqi thinking seems to have been based on the following elements:

• A belief that the Iranian military was so disorganized and demoralized in the wake of the revolution that it would no longer be capable of resisting a determined military attack;

• Interest in altering the terms of the 1975 Iran-Iraq border agreement to reestablish Iraqi sovereignty over the Shatt al-Arab, as well as regaining Arab control over the southern Gulf islands of Abu Musa and the Tunbs that Iran had occupied in 1971;

• Longstanding Iraqi claims on Khuzistan (which Iraq officially described as "Arabistan") and an apparent conviction that the Arab population of that territory would welcome "liberation" by Iraq;

• Apparent belief (probably reinforced by Iranian exiles and opposition elements) that Khomeini's rule would be unable to survive what was expected to be a lightning military defeat and that a successor regime would be composed of individuals less hostile to the existing order;

• Expectation that a quick and total defeat of Iran would shift the balance of power in the Persian Gulf, fulfilling Iraq's ambition to be regarded as a regional superpower and as a leader in Arab politics.

Ironically, the results of the Iraqi invasion produced a set of results precisely the opposite of those intended. The attack helped Khomeini to consolidate his control by rallying nationalist sentiments around the revolution, suppressing internal critics, and accelerating efforts to re-build an effective military machine along Islamic lines. The Arab pop-ulation of Khuzistan resisted the Iraqi advance. Iraq's military offen-sive stalled by November 1980 as Iranian resistance stiffened. As Iran began to counterattack effectively, slowly driving Iraqi forces back to-ward the border, Iraq's great gamble was widely perceived as a fail-ure, undermining its regional influence and leaving it far more de-pendent on the financial and political support of its oil-rich Arab neighbors than ever before. . . .

# Iran on the Offensive

After the Iraqi attack bogged down in November 1980, a military stalemate ensued until the summer of 1981. From September 1981 through May 1982, Iran conducted three major military offensives that forced Iraq to withdraw to the original border in most places. As Iran's forces approached the border, there was a pause that for the first time seemed to offer opportunities for a negotiated settlement. Conse-quently, early 1982 was a time of intensive diplomatic contacts and at-tempted mediation efforts. . . .

Within Iran itself, an intense debate raged about whether to stop at the border or to press its military advantage with an attack into Iraq. In the end, the hardliners won the day. Immediately following the Is-raeli invasion of Lebanon in early June 1982, Iran announced that its forces were "going to liberate Jerusalem, passing through [the holy city of] Karbala" in Iraq. Shortly thereafter, Iran launched the first of a series of massive offensives intended to break through Iraqi defenses, cut Iraqi supply lines between the south and the capital, and bring down the regime of Saddam Hussein.

Although the Iranians never spelled out the underlying reasons for this decision, it appears to have been a mirror image of the original Iraqi decision to launch its invasion of Khuzistan. Carried away by its own revolutionary hubris, Iran seems to have calculated that the Iraqi military was demoralized and would collapse in the face of a determined attack, that the Shi'i population of southern Iraq would welcome the Iranian army as liberators, that the Iraqi regime would dissolve, and that Iran would emerge as the major power in the Gulf.

The outcome was similarly disastrous. The Iraqi army stiffened in the defense of its homeland, and the conflict quickly degenerated into a war of attrition. In the following years, Iraq began attacking civilian targets in Iran with missiles and aircraft (the "war of the cities"), started missile attacks against Iranian oil shipments (the "tanker war"), and eventually resorted to chemical weapons and poison gas to thwart Iran's massed infantry tactics.

If Iran had chosen to sue for peace in mid-1982, it would have been in a good position to influence the terms of a settlement. At that time, Iran was widely perceived as having snatched victory from the jaws of defeat, and its military forces were regarded as perhaps the most potent in the region. By pursuing peace, Iran could have gone far toward restoring its image with both the regional states and the international community, and it could have established a role for itself as a power broker in the region. Instead, Iran once again chose to let its revolutionary fervor overcome a realistic appraisal of its own long-term interests. . . .

# The Role of the United States

The policies of the United States throughout the war were equivocal and often contradictory. In the early stages of the war, the United States was among those nations that called for a withdrawal of Iraqi forces, but this position was never pressed at the UN Security Council. Instead, as the war settled into a stalemate, the United States seemed content to accept the international conventional wisdom to keep hands off and let the two sides batter each other.

That attitude began to change after 1982, as Iran went on the offensive and it appeared that Iraq might be defeated and as the tanker war began to inflict costs on the United States and other nonparticipants. Initially the United States responded to these events by moving closer to Iraq: restoring relations in late 1984, providing Iraq with military intelligence derived from satellites and AWACS [airborne warning and control system] reconnaissance, and launching Operation Staunch which was intended to stop the flow of military supplies to Iran. In 1985, however, in an effort to free US hostages in Lebanon and pursue a "strategic opening" to Iran, the United States and Israel began covertly supplying arms to Iran.

When this operation was revealed in late 1986, it not only caused a sensation in the United States but also cast severe doubts on the credibility of all US policies in the Gulf. After a prolonged period of disarray, the United States began to rebuild its tattered relations with the

Gulf states in 1987 by intervening more actively on the side of the Arabs and Iraq in the war. Eleven Kuwaiti tankers were permitted to register under the American flag, and US forces in the Gulf were expanded to provide convoy protection against Iranian attacks. In response to mining and missile incidents, US forces directly attacked Iranian ships and oil platforms in October 1987 and April 1988, and US rules of engagement were gradually expanded to permit assistance to virtually any ship attacked by Iran.

Assistant Secretary of State for Near Eastern Affairs Richard Murphy visited Iraq in early 1987 and met with Saddam Hussein on May 11. Murphy reportedly promised Saddam that the United States would lead an effort in the UN Security Council for a resolution calling for a mandatory halt of arms shipments to Iran. According to these reports, the UN resolution would not name Iran directly. Rather, it would first call on Iran and Iraq to agree to a cease-fire and to withdraw their forces to the international boundaries. Then "enforcement measures," such as a worldwide arms embargo, would be imposed on the party that rejected the demand. Iran was expected to reject, Iraq to accept. Over the following months, Murphy's pledge was to become a dominant factor in US policy-making on the war.

Six days after Murphy's visit to Baghdad, the USS *Stark* was struck by Iraqi missiles. That event dramatized the military threat in the Persian Gulf and effectively silenced congressional critics who had been resisting a US naval buildup in the Gulf. The irony of the United States responding to an Iraqi attack by virtually declaring war on Iran was not lost on some observers, but it was soon forgotten in the flood of reports about ship movements, Iranian missile emplacements, and an upsurge of Iranian gunboat and mine attacks on neutral shipping. . . .

# Final Straws

Throughout the spring and summer of 1988, evidence accumulated of growing factional disputes within the Iranian leadership. Elections for the third Majlis [Iranian parliament] in early April were extremely contentious. In the days immediately preceding the election, a Kuwaiti airliner was hijacked to [the Iranian city] Mashhad, severely embarrassing those elements of the leadership who were attempting to cleanse Iran's image as a "terrorist state." Several days later, mines again appeared in the central Persian Gulf, one of which struck the USS *Samuel B. Roberts* and set off a new round of clashes with US forces. [Speaker for the Iranian parliament Ali Akbar] Hashemi-Rafsanjani characterized this incident as "An accident . . . which appears to be rigged by elements we cannot yet identify."

This was only the beginning of a series of blows that Iran experienced over a period of three months. Iraq went on the offensive against Iran's disorganized and disheartened military forces, recaptured the Faw peninsula in a lightning attack on April 18, then proceeded to push back Iranian forces all along the front. In mid-May, Iraq carried out a devastating attack on the Iranian oil-transfer site at Larak Island in the southern gulf, destroying five ships, including the world's largest supertanker. Anti-war sentiment began to appear openly in demonstrations in major Iranian cities and, most disturbing of all for the divided leadership, persuasive evidence began to accumulate that Khomeini was severely ill and virtually incapacitated.

Iran desperately attempted to stem the tide, appointing Hashemi-Rafsanjani as the acting commander in chief in an effort to halt the disarray and disintegration of the armed forces and starting a new peace offensive at the United Nations. This was interrupted, however, on July 3 by the tragic downing of a commercial Iranian [Airbus] aircraft by the USS *Vincennes*, killing all 290 passengers and crew.

This terrible accident, coming at the end of a seemingly endless series of defeats, underscored the despair of Iran's position. Despite the enormity of the mistake, Iran was unable to muster sufficient support at the UN to condemn the US action. Its isolation and weakness were never more apparent. As Hashemi-Rafsanjani noted just before the Airbus incident, "We created enemies for ourselves [in the international community]. . . . We have not spent enough time seeing that they become friends."

## Iran Defeated

On July 18, Iranian Foreign Minister Ali Akbar Vilayati sent a letter to UN Secretary General Perez de Cuellar formally accepting Resolution 598 [a cease-fire proposal drafted by the UN Security Council]. Although Iran did not spell out the reasons for this decision, the key factor was probably the changed conditions at the battlefront. Iran had always balked at withdrawing its forces from Iraqi territory without some quid pro quo. That consideration had now been rendered moot by Iraq's recapture of virtually all of its own territory. Resolution 598 had originally been written to favor Iraq, which in mid-1987 was perceived as in danger of losing the war. With the change of fortune on the battlefield, the resolution now offered the prospect of international support and protection for Iran in the face of an effective and determined Iraqi offensive.

Iraq was taken by surprise and initially resisted accepting a cease-fire while continuing its mopping-up operations. Iraq also continued to demonstrate a contemptuous disregard for the Security Council and for world opinion on the use of chemical weapons. A UN investigative team reported to the Security Council on August 1 that "chemical weapons continue to be used on an intensive scale" by Iraq. Only hours later, Iraq launched a massive chemical bombing attack on the Iranian town of Oshnoviyeh. As international pressure mounted, however, Saddam Hussein finally agreed on August 6 to accept a cease-fire on the condition that it would be followed immediately by direct talks. A UN observer force was rushed to the region, and a cease-fire went into effect on August 20. Formal talks began in Geneva on August 25, under the aegis of the UN secretary general. Although the war was not over, for the first time in eight years fighting was suspended.

# The Rise of Democratic Reform

## By Ray Takeyh

*The government of the Islamic Republic of Iran is composed of both democratically elected officials and unelected Muslim clerics. After Mohammad Khatami was elected president in 1997, the contradictions of the theocratic system became increasingly apparent as Khatami and his allies sought to introduce reformist policies. Despite some initial progress, the reform camp was unable to greatly change policies because the conservatives blocked their actions by using their superior legal powers and taking illegal actions. In the following article, Ray Takeyh argues that instead of stalling the reform movement, the conservative backlash has only served to push reformists to adopt more aggressive tactics. Students are increasingly protesting in the streets and demanding democratic reforms. Many conservatives now believe they must find ways to adopt some reforms while retaining the Islamic foundation of the republic. Takeyh is a professor at the National Defense University and a fellow at the Washington Institute for Near East Policy. In 2004 he pub-*
lished The Receding Shadow of the Prophet: The Rise and Fall of Radical Political Islam.

The Islamic Republic of Iran is a regime of paradoxes. The revolutionaries of 1979 sought to usher in a virtuous order in which temporal affairs would conform to divine mandates. Thus, the supreme leader (*Vali-ye Faqih*) was invested with the power to abrogate election results and to select the heads of the armed forces, the judiciary, and the Revolutionary Guards. The dominance of the clerical estate over national affairs was further strengthened by the creation of the Council of Guardians (*Shura-ye Negahban*), which is largely made up of clerics responsive to the dictates of the supreme leader and empowered to screen all candidates for public office and to scrutinize parliamentary legislation for conformity to religious strictures.

Ray Takeyh, "Iran's Emerging National Compact," *World Policy Journal*, Fall 2002, pp. 44–48.

However, the public that had overthrown the formidable monarchy could not be categorically excluded from the deliberations of the state. Thus, the president, Parliament, and local councils were to be chosen by the electorate. Despite the impressive array of powers granted to the clerical oligarchy, Iran's revolutionaries created a governing arrangement whereby collective will would remain an important source of legitimacy. As Iran's clerics were to discover, institutional power devoid of popular legitimacy cannot be sustained over a prolonged period of time. For the theocracy to function, indeed to survive, it had to find a balance between divine authority and popular representation.

# Iran After Khomeini

During the first decade of the Islamic Republic, the unchallenged authority and charisma of Ayatollah Ruhollah Khomeini obscured the regime's underlying contradictions. Iran's contending political factions accepted Khomeini's fiats, while elections were occasions for the public to endorse the ayatollah's candidates. The divisions within the clerical community, where many traditionalist clerics had long viewed actual assumption of temporal power as inconsistent with Shiite theology, went unaired. In the meantime, the democratic promises of the constitution remained largely unfulfilled, as Khomeini neither tolerated dissent nor honored the constitutional pledge of political freedom. Revolutionary excess and rigid dogma became the twin pillars of Iranian politics and, over time, the bond between the regime and the populace gradually began to erode.

In the late 1980s, two events altered the dynamics and nature of Iran's polity. First, the long war with Iraq that had begun eight years earlier ended in 1988, and the cease-fire that followed revitalized political consciousness on the part of a public anticipating some tangible reward for its profound and protracted suffering. Then, less than a year later came the passing of Ayatollah Khomeini. The death of the founder of the Islamic Republic eroded the fragile political consensus and deprived the clerical establishment of both its charismatic leader and its institutional coherence. The ensuing struggle for predominance among the ruling clerics came at a time when the expectations of the public were on the rise. Over the course of the next decade, simmering public discontent and revisionism within the ranks of the clergy and the intelligentsia nurtured a new political movement in Iran.

Not unlike the revolution that it began to critique, the reform movement took shape in universities, seminaries, literary groups, and professional associations. Intellectuals and political activists began to discuss ways of broadening political representation within the context of Islamic governance. Their outlook was informed by their participation as students in the revolution and their subsequent service in the Islamic government, and by their position on the periphery of political viability under the increasingly authoritarian administration of President Akbar Hashemi Rafsanjani. . . .

# Khatami's Rise to Power

From the outset, Iran's seminaries harbored a cadre of clerics uneasy about the direction of the revolution and the growing estrangement of the populace from the religious establishment. A younger generation of clerics—many of whom were linked to Ayatollah Ali Montazeri, ousted in 1987 as Khomeini's heir apparent—sought to interpret Islam in a manner that accommodated popular sovereignty and democratic representation. Hojjat-ol-eslam Mohsen Khadivar, a leading voice for Islamic reformation, captured this sentiment by stressing, "I believe democracy and Islam are compatible. But a religious state is possible only when it is elected and governed by people. And the governing of the country should not be necessarily in the hands of the clergy."

The collaboration of the clerics and intellectuals fused disparate interests within a broad-based movement articulating democratic demands in the language of a familiar faith—yet another parallel with the mobilization leading up to the 1979 revolution. The children of the revolution had come to see themselves as agents of change as opposed to passive pawns in the Kingdom of God. In political terms, this translated into an insistence that the public was the ultimate arbiter of proper governance and that the collective will was the primary source of legitimacy. Iranians could and should shape the ideals and direction of the state through participation in elections and public affairs. Advocacy of this imperative emerged as the guiding principle of the reform movement. The popular appetite for change meant that the reformers had a ready audience and that—unlike previously in Iran's long and troubled political development—the intellectual impulse toward democracy would find fertile ground in which to germinate.

Into this charged arena stepped Hojjat-ol-eslam Seyyed Mohammad Khatami, a mid-level cleric with impeccable revolutionary credentials who was recruited to stand in the 1997 presidential elections in token opposition to the establishment candidate. Despite his position in the clerical establishment, Khatami had long distinguished himself from it, both in his politics and his intellectual enterprises. In 1992, the future president, who was then minister of culture, broke with the Rafsanjani administration over his liberal tendencies and his willingness to grant license for publications and plays that defied the strictures of the regime. After he was ousted, he immersed himself in Western philosophy as a complement to his Islamic training. In his subsequent writings, Khatami dared to contravene the ruling consensus, declaring that "state authority cannot be attained through coercion and dictatorship. Rather it is to be realized through governing according to the law, respecting the rights and empowering people to participate and ensuring their involvement in decision-making." In his campaign speeches, Khatami emphasized the rule of law, the pursuit of justice, and the strengthening of civil society.

This expansive vision of a tolerant Islamic government won the hearts and minds of the Iranian public. And it gained Khatami (who won a whopping 69 percent of the vote) a stunning victory. His election energized the reform movement, as its adherents made the leap from theory to action, from contemplation and critique to accountability and implementation. Now the reformers faced a new challenge: how to navigate the treacherous waters of Iranian politics and institutionalize their ideas. This would prove a much more difficult task than they had imagined.

# Gradual Reform

Prior to 1997, theoreticians of reform had invested considerable time and energy in mapping out a strategy for slowly reclaiming their influence over the course of the Islamic Republic. Having derailed the conservative drive to dominate Iran's tumultuous factional politics with their unexpected capture of the nation's highest elected office, Khatami and his allies now had to put thought into action. However, they faced determined adversaries who rallied their considerable resources to forestall further reformist incursions against entrenched interests.

Once in power, the reformers opted for a strategy of incrementalism, seeking to reform the Islamic Republic gradually from within its own institutions. Taking a dual approach—characterized by the catchphrase, "pressure from below, negotiations from the top"—they sought to respond to burgeoning public demands for greater freedom.

The new president picked his battles carefully and sought to avoid open clashes with the conservatives. To turn up the pressure from below, hundreds of new publications were licensed, censorship guidelines were loosened, and permits for reformist groups and gatherings were issued. The reformers refrained from challenging the wide discretionary power of the supreme leader, which his hard-line allies guarded jealously. Instead, the reformers focused on expanding their institutional power base, taking full advantage of the opportunities accorded by provisions of limited democracy under the Islamic constitution to bring the political competition into the public arena. The reformers sought to buttress their cause by establishing media outlets and creating political parties, notably the Islamic Participation Front.

# The Reformers' Track Record

The strategy of gradualism produced results. Iran's democratic infrastructure was broadened when, in 1998, elections for the constitutionally mandated local councils were held for the first time. Overnight, the number of elected officials in Iran went from 400 to 200,000, with the overwhelming majority of those posts being held by politicians sympathetic to a reform agenda. In the February 2000 parliamentary elections, the reformers captured 189 out of 290 seats, reclaiming an institution that had long served as a bastion of conservative power. At this point, reformers held sway throughout the provincial administrations, in the newly inaugurated municipal councils, and in Parliament. They capitalized on their institutional gains by developing diverse political parties as a means of mobilizing their mass constituency.

The reformist ascendancy produced a number of tangible victories. The most important among these was the reaffirmation of Parliament's prerogative to scrutinize organizations under the supreme leader's jurisdiction. Institutions such as the Ministry of Intelligence, the state broadcasting authority, semi-governmental economic foundations, and even the armed forces were for the first time to be subject to parliamentary oversight. Under its reformist majority, Parliament took up its investigatory license with a vengeance, probing into issues as diverse as the behavior of the security apparatus to the prospect of renewed relations with the United States. Here parliamentary representatives took up the mantle earlier worn by enterprising reformist newspapers.

Khatami's strategy of incremental reform did not lead to the anticipated democratic breakthrough, however. When the reformers began to purge institutions such as the Ministry of Intelligence and to talk about reining in the judiciary, they infringed on the power base of the conservatives. The hard-liners' strategy for retaining the upper hand soon crystallized in the targeting of individual reform leaders, the selective use of violence to intimidate and create division, and the use of the judiciary, along with the Council of Guardians, to block genuine reform. Each time the reformers' inventive circumnavigations of the system managed to gain them even the most ambiguous advantage, they were more than outgunned by the concerted hard-line response, particularly in the courts. Through the cynical use of their institutional powers, the conservatives shuttered hundreds of publications, imprisoned many reformist intellectuals, journalists, and officials, and brutally broke up peaceful student gatherings. The militant faction of the clerical community, the reins of power in its hands, simply refused to countenance a challenge to its anachronistic vision of theocracy.

# A New Phase

Despite the conservative backlash, Iran's reform movement has not crumbled. If anything, it is entering a new and more aggressive phase with the emergence of a younger generation of leaders who are pressing for more immediate results. Among the rising stars of this more robust reform movement are the parliamentarian Mohammad Reza Khatami, who is the president's brother, the dissident cleric Mohsen Khadivar, and student leaders Ali Afshari and Akbar Mohammadi. These reformers have rejected the strategy of incremental change and opted for the more assertive policies of disengagement and confrontation. Some reform parliamentarians and public officials are threatening to leave the government in a move to delegitimize the Islamic Republic, whose survival requires a degree of popular consent. And student organizations are increasingly engaging in active street protests in defiance of the theocracy's prohibitions. Thus, while the reform movement's objectives remain the same—to create a polity that harmonizes religious injunctions with democratic imperatives—its tactics are evolving. Instead of changing the system from within, reformers are increasingly seeking to exert pressure from without.

The new reform strategy crystallized this past summer [2002] when the Islamic Participation Front, Iran's largest reform party, which controls 130 out of 290 parliamentary seats and 5 ministries, warned that if the conservatives continued their obstructionism, it would disengage from politics. Ebrahim Yazdi, one of Iran's most prominent dissidents, pointedly declared, "We are approaching a turning point. Basically, down deep, there is confrontation between tradition and modernity."

University students, chafing under suffocating cultural restrictions and frustrated by the lack of employment opportunities, are increasingly taking to the streets in protest. At a time when the regime is incapable of providing employment for half of the 800,000 people who enter the job market each year, and with the "moral police" once more cracking down on those who deviate from religious codes of dress and conduct, the young are demanding radical change. In defiance of government bans, student associations, such as the Office for Consolidation of Unity and the Union of Islamic Students, have emerged as the vanguard of the newly emboldened reform movement. A protesting student captured the spirit of the new partisans of change in declaring, "We aren't afraid. They can't frighten us."

# New Strategies

The Internet has become a potent source of information in the place of the banned newspapers, and a means of organizing. Many Iranian journalists are establishing websites. "Technology always wins, and therefore the closure of reformists' newspapers is unimportant when there is the Internet," declares one writer.

The conservatives' strategy has backfired. Their policy of obstructing evolutionary change and Khatami's strategy of cohabitation has led to the rise of an even more determined reform movement whose leaders are not just impatient but capable of mounting a serious challenge to the legitimacy and viability of the Islamic Republic.

As the reform movement alters its tactics, it is forcing the conservatives to shift their perspective. Increasing numbers of conservatives are beginning to appreciate that their long-term relevance is contingent on their ability to engage Iran's youth and on their commitment to the creation of a tolerant society. . . .

The struggles [since 1997] have persuaded an important segment of the conservative bloc that the current stalemate will only lead to the demise of the governing system established in 1979. They know that they must start on the path to reform or face an angry constituency whose demands cannot be ignored indefinitely. The pragmatic forces within the conservative camp appear to be looking for a way to advance the cause of pluralism while retaining the Islamic boundaries of the state. If they succeed, Iran's contending factions may yet confound their critics and fashion a compromise that fulfills the revolutionary pledge of 1979 to create a political order that is both representative and responsive to traditional values.

**CHAPTER 2**

# Iran's Internal Conflicts

# The Hidden Strength of Iran's Regime

By Bijan Khajehpour

*Most analysts divide Iran's ruling establishment into the opposing reformist and conservative camps. However, Iran-based Bijan Khajehpour contends that no fewer than six different factions are important in the present regime. Despite disagreements on political and social liberalization, these factions are united by a commitment to Islamic government and a consensus that national security and the economy are the most pressing issues. In this article, Khajehpour argues that to avoid standoffs among the factions, Iran's leadership will focus its new initiatives on foreign policy and economic reforms, while setting to the side issues of political and cultural reform. Although students continue to protest for strong democratic reform, the regime is not about to collapse. In fact, postrevolution Iran's hidden strength is its adaptability. Bijan Khajehpour is the editor of the* Iran Focus *newsletter.*

On December 7 [2002], Iranian security forces and members of the hard-line conservative Baseej militia reportedly attacked a crowd of 10,000 demonstrating in solidarity with the students outside the Tehran University gates, while leaving the 2,500-strong campus rally unmolested. Smaller crowds kept up the demonstrations in succeeding days, amid further Baseej assaults. The student-led protests, this time sparked by the deeply unpopular death sentence handed to history professor Hashem Aghajari for his comments critical of Iran's clerical establishment, have reasserted a role for Iranian youth in the complex and tumultuous struggles unfolding over the political future of the Islamic Republic.

A major problem in Western media coverage of these struggles is their over-simplification into a clichéd confrontation between "reformists" and "conservatives." This paradigm fails to explain the current state of affairs in Iran. Certainly, divisions in Iran's ruling establishment were greatly sharpened by the resounding victories of President Mohammed Khatami, and the "reformist" Second of Khordad Front loosely associated with him, in the elections of 1997 and 2001. The judiciary in particular has resisted and repressed developments like the lively pro-reform press that flourished after Khatami's first electoral triumph. But the divisions in the Iranian regime do not break down neatly along reformist and conservative lines. In fact, one of the key bottlenecks in Iran's political development lies in the fact that the factions favoring political liberalization are at odds with economic liberalization and the faction that promotes privatization and free-market economics is distant from views such as political pluralism.

The resulting welter of deadlocks and compromises has frustrated those inside and outside Iran who hoped that the self-styled reformist politicians would bring a rapid and comprehensive transformation of the Iranian polity. But the very need of the multiple factions to build consensus means that Iran's leaders are not necessarily "sitting on a time bomb," as a recent *Newsweek* headline posited.

## An "Exhausted" Society

One should remember that it was the Iranian people—specifically, the electorate—that injected the current of reformism into Iranian politics in 1997 and has kept it alive until this day. At present, however, it is more and more obvious that Iranian society has embarked on a path of depoliticization. The difficulties faced by the reform movement in the past few years have disillusioned the people, disinclining them to remain involved with the country's political dynamics. Factional infighting has "exhausted" society, in the words of parliamentary speaker Mehdi Karroubi, one of the most outspoken leaders of the Second of Khordad Front. The majority of the Iranian middle class—the group Western observers normally expect to spearhead social change—has a direct interest in political stability, due to opportunities created by fast-paced urbanization, investment schemes attractive to small investors and the growth of small and medium-size private-sector industries.

Despite its apparent dissatisfaction with the slow pace of reform in the country's political structure, Iranian society does not appear bent on forcing change at this stage, favoring instead a gradual process. It remains to be seen whether popular disgruntlement with reformist politicians will reduce participation in the parliamentary elections approaching in February 2004[1] or the presidential elections of 2005. But it is clear that the perceptions of Iranians under 30—who constitute 70 percent of the population—will be crucial in determining the turnout. Therefore, the country's numerous political factions are under pressure to respond somehow to the growing needs of Iranian youth.

## Interlocking Factions

Rather than two diametrically opposed camps, one can distinguish six different currents present within the ruling establishment. A hard-line left, composed of labor unions and other groups with a revolutionary ideology, is mainly concerned with the social justice values of the Islamic Revolution of 1979. These forces have successfully guarded early revolutionary policies such as subsidies for staple goods and worker-friendly labor laws. "Ultra-reformists" led by Saeed Hajjarian and Reza Khatami focus chiefly on political freedoms and democratization. Emerging from the hard-line left is a moderate left that has revised its views and become the main force behind political and cultural reform. This group's champion is President Khatami.

On the right, a group of technocrats has developed a free-market economic vision in the 1990s, based on a more liberal political framework. The key figure in this current remains former President Akbar Hashemi-Rafsanjani. Ayatollah Ruhollah Khomeini's successor as supreme leader, Ayatollah Ali Khamenei, is at the head of a group of clerics known as the "cultural conservatives," who concern themselves mainly with cultural policies rather than economics. This group has relaxed its conservatism somewhat based on the actual experiences of the Islamic Republic. Finally, hard-line conservatives rally around the monopolistic view—brooking no compromise on clerical control over all aspects of politics and society—that unites arch-conservative clerics and interest groups inside and outside the ruling establishment.

Though the balance of power among these sometimes interlocking factions is constantly changing, the experience of the past five years has shown that none of them can efficiently control the political affairs of the country from the parliament. Managing Iran has always required compromise solutions between the key stakeholders. For this reason, supra-parliamentary bodies such as the Supreme National Security Council [charged with "preserving the Islamic Revolution"] and the Expediency Council [the Supreme Leader's advisory board] where

---

1. In the February 2004 elections, unelected religious conservatives barred about 2,500 reformist candidates from running for seats in parliament. Voter turnout was a record low 50.57 percent— but apparently not due to popular discontent with reformists. In fact, many reformists called for a voter boycott, hoping a low turnout would discredit the result and put pressure on conservatives, who ended up winning the majority of the seats.

representatives of all major factions hold seats, carry heavy weight in the post-revolutionary political structure. Not all the deliberations of these bodies are riven by conflict.

## Consensus and Conflict

Wide-ranging consensus prevails among the key factions that national security is the top priority for today's Iran, especially given current regional tensions. The concept of a "national interest" in security matters, wholly detached from the ideological considerations that animate debates on other matters, has emerged. Iranian policymakers agree that the country should concentrate on keeping good relations with its Middle Eastern neighbors, as well as with Europe and Asia. Based on a consensus among top decision-makers, Iran and the European Union are about to start negotiations toward a trade and cooperation agreement which also provides for talks on political development and human rights. By contrast, there is continued debate and disagreement on the necessity and future of relations with the United States, meaning that confusing messages about Iranian-US ties are sent out from Iran. For the time being, few deviate from the official line that the US needs to prove its good will before Tehran will make overtures of its own. . . .

Consensus also holds that economic conditions need to be improved and jobs created—hence, the current focus of state institutions on reforming the economy. Privatization and increased efficiency in the state sector will be among the key programs and to enhance economic performance and also to reduce the scope for corruption.

But there continue to be major disagreements over how economic reform can be achieved, in light of Iran's revolutionary commitment to "social justice" and the state-centered economic policies of the past two decades. On political and cultural issues, there are sharp divergences among the factions, especially regarding the pace and nature of political reform. These disagreements were less significant as long as the nation was unified by revolution, war with Iraq and post-war reconstruction. But in the 1990s, it became clear that new rules of the game were required.

## The Chinese Model

Khatami's initial strategy to find a common denominator for the political factions was to state that the constitution laid out the political agenda for Iran. But the stalled progress of reform over the past few years has underlined the inadequacy of this strategy. The unelected Council of Guardians [a board of twelve jurists with the authority to veto laws passed by Parliament if found incompatible with Islamic law] retains the ability to bar candidates from running for election and to reject legislation passed by Parliament, while the president lacks the authority to enforce the constitution's provisions. In response, in September 2002 Khatami presented to Parliament the so-called twin bills which address precisely these two issues—the first would curb the powers of the Council of Guardians, while the second would enhance presidential powers. Should these bills become law, Iran's political system would enter a new stage in which the president and the supreme leader, who can intervene in all manner of judicial and legislative decisions, would enjoy a dual sharing of power. Interestingly, Supreme Leader Ayatollah Khamenei recently called the constitution the main charter of the Islamic Republic, and advised state officials to use "legal channels" to settle their disputes. These "legal channels," however, would remain limited should Khatami's bills be blocked by the Council of Guardians or the Expediency Council.

Given this standoff, for the time being Iran's top leaders feel they have no choice but to stick to a Chinese model of reform: Iran's new initiatives will remain in the areas of foreign policy and economic policy, while political and cultural reforms will stay on the back burner. The Chinese model seems to have the blessing of Khamenei, Hashemi-Rafsanjani, Khatami and Karroubi alike, though the latter two have some reservations.

Two important factions disapprove of the Chinese model: the ultra-reformists and the hard-line conservatives. Clearly, the former faction is unhappy that the hoped-for political liberalization of the country does not seem feasible at present. Their discontent is reflected in the student demonstrations and other protests, as well as in their calls for Khatami's resignation. The hard-line conservatives are opposed to any type of reform that could sabotage their position. Economic reforms such as breaking of monopolies, liberalizing imports, promoting foreign investment, unifying the exchange rates and increasing controls on smuggling have hurt vested interests that back the hard-liners.

## The Republic Is Not About to Collapse

The centrist consensus behind the Chinese model and corresponding opposition from the factions furthest left and furthest right will characterize Iran's political scene for some time to come. While the centrist consensus may try to further marginalize the opposing currents, for the time being the balance of power is such that the ultra-reformist and hard-line conservative forces will have the ability to undermine the political process. . . .

Meanwhile, the forces on both ends of the political spectrum are likely to push forward with their agendas. Peaceful student protests against the lack of political liberalization will proceed, even as interest groups supporting the hard-liners place more obstacles in the path of reform. The student protests, while dramatic and important for long-term political consciousness, do not appear destined to turn into a mass popular uprising, due to Iranian society's preference for evolutionary change. Political reforms will only be enacted at a sluggish pace, as they require consensus-building efforts on a larger scale than does privatization of the economy. The consolidation of the centrist coalition could pave the way for quicker, though limited, progress.

As student protests and government crackdowns continue, some advance the weak analysis that the Islamic Republic of Iran is about to collapse. The Islamic Republic is a highly adaptable regime which has matured over the past decade. Iran's failure to evolve into a Western-style democracy does not mean that the current regime is not sustainable. Indeed, the backbone of the post-revolutionary system's sustainability might be the fact that it continuously looks more fragile than it really is.

# The Struggle for Women's Rights in Iran

## By Ziba Mir-Hosseini

*The issue of hejab, the complete covering of women's bodies with the exception of the face and hands, has long encapsulated the struggle over women's rights in Iran. In 1936, Mohammad Reza Shah banned hejab, which remained illegal until the Islamic Republic made it compulsory in 1983. Although official debate over hejab was conspicuously absent in the following years, the emergence of the reform movement in 1997 encouraged public discussion not only of this issue, but of women's rights generally. In the following selection, Ziba Mir-Hosseini argues that despite attempts by conservatives to maintain traditional gender inequalities, the struggle over women's rights has entered a new phase, forcing conservatives to compromise on many issues. Mir-Hosseini is a research associate at the Centre for Near and Middle Eastern Studies at the University of London. She is the author of* Islam and Gender: The Religious Debate in Contemporary Iran *and* Feminism and the Islamic Republic: Dialogues with the Ulema. *She also directed the award-winning documentary* Divorce Iranian Style.

Women's votes were decisive in the outcome of the 1997 presidential election, in which [Mohammad] Khatami defeated [Ali Akbar] Nateq-Nuri, the candidate who enjoyed the support of the clerical establishment. During his campaign, Khatami frequently spoke of youth and women as two groups whose aspirations and demands must be acknowledged and validated by the government. His language was novel in the political climate of the day. Unlike his fellow clerics, who continued to speak in the *language of duty* (*taklif*), Khatami spoke in the *language of right* (*haqq*). It was this that made his address and

Ziba Mir-Hosseini, "The Conservative-Reformist Conflict over Women's Rights in Iran," *International Journal of Politics, Culture, and Society*, vol. 16, Fall 2002, pp. 39–40. Copyright © 2002 by Human Science Press. Reproduced by permission of the publisher and the author.

campaign appealing to women and the youth. He also broke other unspoken clerical codes of public conduct. For instance, in an interview with *Zanan*, the women's magazine with an Islamic feminist perspective, Khatami spoke about his relationship with his wife and his children and came across as a liberal Muslim. The impact of his smiling face on the cover of the magazine and his willingness to answer its provocative questions contrasted sharply with the refusal of the establishment candidate, Nateq-Nuri, to answer another list of questions, published in the same issue.

# Silencing the Call for Rights

Soon after Khatami took office, the working relationship forged in the previous decade between Parliament (*Majles*) and clerics of opposing political tendencies, in the form of a fragile consensus on the need to reform Islamic laws related to women and family, started to break down. This consensus had seen the enactment of a number of measures redressing some of the gender discriminations introduced soon after the Revolution. For instance, restrictions on subjects that women could study were removed (1986); family planning and contraception became freely available (1988); divorce laws were amended so as to curtail men's right to divorce and to compensate women in the face of it (1992), and women were appointed as advisory judges (1992). Women members of the *Majles* had acted as a strong lobby in securing the cooperation of the seminaries and eminent clerics and pushing for such legislation. All this changed after Khatami took office, when the Conservatives lost their absolute control of government. Fearing that popular support for the new language of rights—as embodied in the slogans "civil society," "freedom of expression" and "rule of law"—would further undermine their position, the Conservatives tried to reassert their authority, counting on institutions still under their control to block and frustrate every move of the Reformist government. The fifth *Majles* (1996–2000), in which the Conservatives still had a majority, was the first to intervene. The newly-created *Majles* "Women's Commission," whose members until then had united in defense of women's rights, proposed two bills that became infamous for their antiwomen slant. Both bills were part of a concerted Conservative effort to frustrate the reforms promised by Khatami by using his own slogan, "the rule of law." Their main purpose was to put the lid on debates that were emerging in the burgeoning Reformist press.

The first bill, "Adaptation of Medical Services to Religious Law," extended the imposition of *shari'a* [Islamic law] codes of gender segregation to medicine, a realm that had been left more or less untouched since the Revolution. The second, "Banning the Exploitation of

Women's Images and the Creation of Conflicts between Men and Women by Propagating Women's Rights outside the Legal and Islamic Framework," sought to prohibit the lively press debate on women's rights as well as press coverage of the dynamic film industry.

Both bills became laws in July 1998, but they were so far removed from the realities of Iranian society that they proved impossible to implement. . . .

## The Issue of *Hejab*

In sharp contrast to other areas of women's rights, there had been no meaningful debate on the issue of *hejab* [Islamic female covering]. Even an outspoken cleric such as [Hojjat ol-Eslam Seyyed Mohsen] Sa'idzadeh and a daring magazine like *Zanan* had been careful to avoid the *hejab* issue in their critique of the construction of gender rights in Islamic law. In practice, however, many women had challenged the imposition of *hejab* from the outset, and its frontiers were constantly pushed back—especially by the very generation of women who were born after the Revolution and indoctrinated to accept *hejab* as a supreme value.

This paradox, that is the public silence over the issue of *hejab* and the defiance of it in practice, has its roots in the ideology and policies of the *ancien régime*. Reza Shah, the first Pahlavi monarch, promoted a policy of "unveiling" (*kashf-e hejab*) as part of his modernization campaign. In 1936 a law made wearing *chador* [an all-enveloping black garment worn by some Muslim women] or any head-covering apart from a European hat in public an offence, and women wearing traditional Iranian cover were arrested and their cover was forcefully removed. This outraged not only the clerical establishment and religious families, who saw the law as a direct assault on an Islamic mandate, but also many ordinary women for whom appearing in public without their *chador* was tantamount to nakedness. Others, both men and women, welcomed the law as a first step in granting women their rights. Since then, the *hejab* issue has become a deep wound in the Iranian political psyche, arousing strong emotions and reactions on all sides. It is a major arena of conflict between the forces of modernity and Islamic authenticity, where each side has projected its own vision of morality.

In the early years of the Revolution, *hejab* once again became a divisive issue. Many women strongly resisted attempts by the revolutionary regime to make *hejab* compulsory. But it was to no avail, and compulsory *hejab* was legally enacted. In 1983, Article 102 of Islamic Punishments (*ta'zirat*) made appearing in public without *hejab* an offence against public morality, punishable by the "Islamic" penalty of up to seventy-four lashes. The imposition of *hejab* was defended and enforced with such vigor in those years that it gradually became one of the cornerstones of the Islamic Republic. In their Friday sermons, their lectures, and their writings, political clerics often spoke of the success and the authority of the Islamic Republic in terms of its policy of compulsory *hejab*.[1]

But after 1997, with the opening of political space and the expansion of the discourse of rights, debate over the issue of *hejab*, successfully suppressed since the Revolution, resurfaced. A number of articles appeared in the Reformist press, questioning the wisdom of the policy of imposing *hejab* and pointing out its anachronistic incompatibility with the new discourse advocated by the Reformist government. For instance, in June 1999, *Neshat* [a Reformist newspaper] featured two daring articles which were occasioned by the case of Merve Kavakci, an Islamist deputy in the Turkish parliament, who was forbidden to take her seat because of her insistence on observing *hejab*. The case was widely debated in Iranian press, with the Conservative papers using it as a sign of the lack of freedom in Turkey for women to fulfil their religious duties. Fa'ezeh Hashemi, a deputy in the fifth *Majles* and daughter of former President Hashemi Rafsanjani, wrote a letter to Kavakci offering her support on behalf of Iranian women. Kavakci rebuffed the initiative, saying that she did not need the support of those who themselves imposed *hejab* and did not respect the democratic rights of women. Hassan Yusefi Eshkevari, a cleric, in the first *Neshat* article, "Defending Values with the Logic of Democracy," after exploring the reason for such a rebuff, wrote: "In today's world, at both national and international levels, religious opinions or moral values can be defended with only one logic, and that is the logic of democracy." Eshkevari's point was followed up two weeks later by Farhad Behbahani, who wrote that "the logic of religion in the issue of *hejab*—like other issues—is in line with democracy." He then entered a discussion of the Koranic verses on *hejab* and argued that the whole issue comes under the realm of belief; nowhere does the Koran speak of it as an injunction. In Behbahani's words: "Not observing *hejab* merely constitutes a 'sin,' answerable only to God, and is different from a 'crime' which is a social matter and can be dealt with by those in charge in society.". . .

---

1. The mandatory wearing of *hejab* has been part of Iran's legal code since 1980. The penalty for noncompliance ranges from a considerable fine to imprisonment, although enforcement varies. In rural areas, women commonly wear the *chador*, a black bell-shaped garment that covers the entire body, while in urban centers such as Tehran, women wear long coats, colored head scarves, and even makeup. In a recent interview, President Mohammad Khatami has suggested that women should have the right not to wear the veil.

# The Battle Continues

With the opening of the Sixth *Majles* in June 2000, a new front opened in the battle between Reformists and Conservatives. The latter—this time using the Guardian Council—have so far managed to frustrate most legislative moves of the Sixth *Majles*, called the *Majles* of Reforms (*Majles-e Eslahat*), but not with impunity. In September [2000] the *Majles* ratified a bill to eliminate discrimination between female and male students in regard to scholarships for study abroad. This—the first bill of its kind that the *Majles* Reformists brought to honour their preelection promise to do away with gender inequalities in law—put them into direct confrontation with the clerical establishment. It was condemned by Friday prayer leaders, and Ayatollah Makarrem-Shirazi, a high-ranking cleric in Qom, wrote an open letter to the *Majles* demanding withdrawal of the bill. The letter, read out loud by a Conservative deputy, accused the *Majles* deputies of undermining Islam and people's honour (*namus*) for the sake of votes. The deputies ignored the letter and ratified the bill, but it was rejected by the supervisory Guardian Council as "non-Islamic." The Council offered no justification for its rejection, but it could no longer ignore the popular impact of its decisions, for which it blamed the *Majles*. Ayatollah Meshkini, the Friday prayer leader in Qom, said it all in a sermon: "Ratification of such a bill by the *Majles* causes antagonism of the youth towards the clerical establishment and the Guardian Council."

The rejection of this bill brought out the anachronism and gender bias in the law, and the whole issue became a topic of debate in the press. Conservatives argued that the bill contradicts the *shari'a*, since in order to study abroad a woman needs the consent of her father or legal guardian. Reformists pointed out that women pursuing higher education cannot be considered minors who need such permission, that the *shari'a* requires the permission of a woman's guardian only for her first marriage, and that there are no grounds for extending this requirement to higher education. A compromise was reached when the *Majles* amended the bill, using the prerevolutionary Passport Law, which allows a woman over eighteen to travel without her father's permission but requires a married woman to have her husband's permission. In March 2001, the Guardian Council approved this amended bill.

Two other bills, raising the minimum age of marriage for women to fifteen and for men to eighteen, and extending women's grounds for divorce, which the Guardian Council had rejected earlier, have gone to the Expediency Council (*Shura-ye Maslehat-e Nezam*), the final arbitration authority in disputes between the *Majles* and the

Guardian Council. It remains to be seen what sort of compromise will be reached.[2]

With the emergence of the Reformist movement in 1997, the struggle over women's rights in the Islamic Republic of Iran entered a new phase. It became part of a larger struggle between Conservatives and Reformists over two notions of Islam. One is an absolutist and legalistic Islam, premised on the notion of "duty," that tolerates no dissent and makes little concession to the people's will and contemporary realities. The other is a pluralistic and tolerant Islam, based on human rights and democratic values. It is perhaps too early to say how and when the notion of Islam as advocated by the Reformists will begin to prevail and to redress the gender inequalities in *shari'a* laws.

---

2. In March 2004 the Expediency Council approved a bill requiring parents to obtain court permission for marriages of girls younger than thirteen and boys younger than fifteen.

# Iran's Youth Will Force Change

## By Bahman Baktiari and Haleh Vaziri

*Young Iranians face a bleak future under the present regime. Cultural repression, economic stagnation, a high unemployment rate, and international isolation have convinced a large proportion of those born after the 1979 revolution that reform of the Islamic Republic is necessary. Though Iran's youth initially supported reformist president Mohammad Khatami, they have become impatient with the slow pace of change and seek to confront the conservatives through pro-reform organizations, mass protests, and student revolts. In the following selection, Bahman Baktiari and Haleh Vaziri argue that despite their commitment to reform the theocratic system, thus far young Iranians have lacked both the consistent platform and united front necessary to mount a sustained campaign that will produce the changes they desire. Bahman Baktiari is director of the international affairs program and associate professor of political science at the University of Maine. He is also the author of* Parliamentary Politics in Revolutionary Iran. *Haleh Vaziri is an activist and political scientist who specializes in human rights and Iran. She is the coauthor of* Safe and Secure: Eliminating Violence Against Women and Girls in Muslim Societies.

In May 1997, Mohammad Khatami surprised his clerical colleagues and international analysts with a landslide win in the Iranian presidential race. Khatami's electoral defeat of the establishment candidate, Ali Akbar Nateq-Nouri, revealed Iranians' expectations that their new president would ease the Islamic Republic's restrictions on cultural and social freedoms in the name of religion, execute the rule of law consistently, and strengthen civil society. Four years later, in June 2001, Iranians reelected Khatami by an even wider margin, showing their patience with the president, who still encounters intense opposi-

Bahman Baktiari and Haleh Vaziri, "Iran's Liberal Revolution?" *Current History*, January 2002.

tion from Islamic Republic conservatives worried about the erosion of clerical supremacy and the advent of secularism.

# Khatami and the Reform Movement

During both presidential campaigns, Khatami appealed to four constituencies: a middle class chafing under Islamic constraints on social life and frustrated by Iran's international economic isolation; intellectuals objecting to the ruling clergy's violations of human rights; women resisting the clerics' efforts to limit their rights and roles in Iranian society; and youths hungry for a loosening of cultural and social restrictions and in need of higher education and jobs.

These constituencies formed the core of the reformist movement, which has captured not only the presidency twice but also majorities in the municipal council elections of February 1999 and in parliamentary races a year later. Reformists in government have not succeeded in fully implementing their platform, however. They have so far been unable to guarantee freedom of expression—the centerpiece of Khatami's domestic agenda—to repair Iran's economy, or to take full advantage of the president's call for a "dialogue of civilization" to normalize relations with the United States. Although popular participation in culture and politics has increased during the Khatami era, the Islamic Republic's conservatives have fought reforms tenaciously. With Supreme Leader Ayatollah Ali Khamenei often on their side, the conservatives control the security forces as well as the judiciary. Launching episodic and unpredictable waves of repression, they have kept the reformists on the defensive.

This stalemate in Iranian politics inspires three questions: Given its victories at the ballot box, why has the reformist movement stalled? How long will Iranians—particularly youths—wait for reforms to have a tangible impact on their lives? Do Khatami's constituents have any alternatives to waiting for the success of reforms?

Arguably, the reformist movement has stumbled over three obstacles. First, the dualism of executive power between a secular and a religious leader renders gridlock more likely in the Islamic Republic's policy making. Furthermore, President Khatami lacks the personal qualities and skills to push reform past conservative opposition; he is a humanist intellectual with a passion for ideas but has chosen to preserve the system in moments of crisis. Thus, in the movement toward reform, Khatami's constituents, especially Iran's young, are out ahead of him. Yet these constituents also have a long way to go before crossing the ever-elusive finish line: formulating policies to delineate the role of religion in public life, to democratize the procedures and substance of Iranian politics, and to engage Iran in a constructive network of foreign relations. . . .

# Reform and Reaction

Significantly, Ayatollah Khamenei disagrees with Khatami's understanding of the Islamic Republic's essence and objectives. Commemorating the twelfth anniversary of [Ayatollah] Khomeini's passing, the supreme leader claimed that his mentor was the "symbol of Islam and the people, because by relying on Islamic sovereignty and power, he founded an Islamic society on the basis of spirituality and justice. . . . Like any person who is acquainted with Islam, the late Imam believed, as we continue to today, that it is Islam and Islamic principles that ensure the happiness, welfare, freedom, and dignity of a nation. It is Islam that ensures that justice is upheld in society. Reliance upon the people, in the true sense of the word, is made possible only in the context of upholding Islam and Islamic principles."

If election results are any indicator, the majority of eligible Iranian voters agree with the president's emphasis on the republican quality of the Islamic Republic. Still, during every election since 1997, the conservatives have done their best to skew the process and outcomes. The Council of Guardians disqualified prominent reformists, especially women, from the parliamentary and presidential races, and the conservatives demanded recounts of votes in localities where they lost by narrow margins. These measures appear to have been counterproductive, however, because the February 2000 parliamentary contest gave the reformists some two-thirds of the seats.

When the newly elected reformist deputies sought to amend restrictions on the press passed by the previous parliament, the conservatives flexed their muscle. Reformist legislators wanted to revoke a law preventing once-banned publications from reopening under new names and barring the reemployment of journalists from these publications. As the reformist majority passed a bill repealing the stipulations, the conservative minority appealed to the supreme leader, who took the unprecedented step of intervening in the legislative process. Ayatollah Khamenei vetoed the more liberal press law on the grounds that its text contradicted the precepts of Islam. The parliamentary majority—the popular will—was trumped by the practice of divine sovereignty.

# The Repression of 1379

During the Persian calendar year 1379 (March 2000–March 2001), as Khatami's first term was ending, Iranians witnessed a campaign of repression unprecedented since Khomeini's death 11 years earlier. Conservative leader Ayatollah Mesbah Yazdi preached that recourse to violence is "obligatory" if the Islamic Republic cannot otherwise defend itself, "even if thousands of people must perish." Devout Muslims must "kill on the spot" anyone who "insults Islam or the Prophet." Ayatollah Khamenei also declared unequivocally that violence was a legitimate tactic to oppose the flouting of Islamic precepts.

The press stood at the front lines of the conservative assault against the reformists. During 10 days in April 2000, the judiciary, headed by Ayatollah Hashemi Shahrudi, closed 16 reformist newspapers without any hearings. Conservatives forced all the pro-press officials out of the Ministry of Islamic Guidance and Culture, with the result that requests for publications were increasingly refused. Between April 2000 and March 2001, roughly 2,000 requests to publish were filed with the Press Supervision Council; only 62 publications received licenses. In March 2000, Iran's newspapers had a total daily circulation of approximately 3,120,000, which dropped to 1,750,000 a year later—a 45 percent decline.

While readers watched proreform newspapers disappear, other members of Iran's fledgling civil society confronted the conservatives' wrath: academics, journalists, lawyers, liberal clerics, publishers, student leaders, and even some midlevel government officials were arrested. Charged with seditious activities, some were sentenced to lengthy jail time following public or closed-door trials, while others, not yet tried, are still held in secret confinement, their whereabouts unknown to their attorneys or families.

Among those arrested was eminent theologian Hassan Yussefi-Eshkevari, who was charged with the crime of apostasy; he defended the principle of separation between mosque and state, and contended that Islam does not require women to veil. Although condemned to death after a closed-door trial, the Court of Appeals overturned Yussefi-Eshkevari's sentence. Mashallah Shamsolvaezin, the editor in chief of several banned dailies, is now serving a 30-month prison term for publishing an article urging the death penalty's abolition in the name of Islam. In April 2001 the judiciary accused and jailed some 40 nationalists for conspiring to overthrow the Islamic Republic. . . .

During this wave of repression, President Khatami was reduced to voicing his disapproval, since the reformers were powerless to protect those whom they had urged to speak freely. Khatami could only protest: "Three and a half years, and still one has not enough authority for the accomplishment of such a grave responsibility!" Without the authority to redress violations of the constitution, some of Khatami's aides suggested that he would not run for reelection.

# Iran's Youth

As the 2001 presidential race neared, reformists worried about the impact of this repression on voters. With more than two-thirds of Iranians under 30, young people form the largest voting bloc. Youths under 30 comprise 52 percent of eligible voters, and those ages 16–20 form 22 percent. Most young people have little or no memory of the Islamic revolution, did not participate in the referendum establishing the Islamic Republic, and did not vote to ratify its constitution in 1979, yet they have dealt with the consequences of these events, which their parents' generation celebrated—at least initially.

Iranians 16 years and older have many reasons to be skeptical of Khatami in particular and to resent the Islamic Republic more generally. Not forgotten are the July 1999 campus riots, sparked when police and right-wing vigilantes attacked a Tehran University dormitory as students protested the closure of a proreform campus newspaper. One student was killed and at least 20 were injured, igniting the most significant unrest in Tehran and other cities since the 1979 revolution.

Students around the country chanted pro-Khatami slogans, but the president cautioned that violence would not expedite reform, which, he said, must come slowly and deliberately. Although at first caught off-guard by the students' demands, the ruling clerics quickly closed ranks. Khamenei and Khatami authorized the suppression and arrest of student rebels while assuring the nation that the entire episode, including police misconduct and vigilantism, would be investigated. Many young people initially felt betrayed as Khatami sided with the preservation of clerical supremacy. Students have since not shied away from politics, joining Islamist and reformist organizations. The largest student group, the proreform Office to Foster Unity, is some 500,000 strong. But even without joining such organizations, the under-30 generation is marked by its political consciousness; nearly every aspect of a young Iranian's daily existence is politicized. Youths lead a double life. Publicly, they must obey the ruling clergy's prohibitions against socializing with the opposite sex, listening to pop music, consuming alcoholic beverages, wearing shorts if a man, using cosmetics if a woman. In the privacy of their own homes, however, they pursue intimate relationships with the opposite sex, enjoy parties where alcoholic drinks flow and Iranian and Western pop music plays, and experiment with dress and cosmetics as a form of self expression (although they continuously fear being caught even behind closed doors).

# A Bleak Future

Young Iranians aspiring to a university education must endure a grueling entrance examination, knowing that their prospects for matriculation are slim. Out of 1.5 million high school seniors who took this exam in 2001, the university system accepted a mere 150,000. University graduates then struggle to find employment. For every 23 graduates, there is one job; 85 percent of Iranians under 25 are unemployed. Without earning any income, young people literally cannot afford to marry, despite the cultural value invested in matrimony.

In the absence of better alternatives and still hopeful that Khatami could improve their lives, young Iranians, two years after the student rebellion, are still the most vocal about how Iran is governed not just systemically, but on a daily basis. Watching Iran's promise thwarted by widespread corruption, high unemployment, global isolation, and repression, youths have demanded accountability and, ultimately, democracy from the reformists. During campaign rallies a week before the 2001 presidential race, students carried placards with such slogans as "Khatami the hero, the hope of the young," "Freedom of thought, always, always," "We will make the crisis-makers despair by voting for Khatami again," "Human beings must be allowed to ask questions," and "We have come to renew our allegiance to reform."

Khatami's reelection, while hailed by young Iranians, provoked the ire of some conservatives. On the order of judiciary chief Ayatollah Shahrudi, security forces launched a wave of public floggings targeting youths. From August to September 2001, over 400 people, most under 25, were flogged publicly, accused of "consuming alcohol, having illicit sex, or harassing women." Outraged, Khatami warned that in a "society where discrimination, poverty and graft abound, one cannot expect youngsters not to break the law. . . . With tough punishments, you cannot remove social corruption." Foreign Minister Kamal Kharrazi also denounced the floggings, concerned about harm to the Islamic Republic's international image.

Conservative clerics applauded the punishment. Chief Justice Ayatollah Mohammad Mohammadi-Gilani even claimed that the supreme leader had approved the floggings, insisting that the offenders "should be beaten to the point where the whip breaks the skin and scars the flesh underneath." Conservatives cared little about the Islamic Republic's image abroad, nor would they listen to domestic critics of the beatings. They believe that flogging is a deterrent to un-Islamic behavior and rising crime rates. Defending this form of punishment, Ayatollah Shahrudi stated, "All should be sensitive toward the issue of the promotion of corrupt means and fight against the enemies' efforts to deprave our children."

# The Soccer Riots

The wave of floggings heightened tensions on the streets of Iran's major cities, where youths have pushed the limits of those cultural and social freedoms that have increased during the Khatami years, despite the conservative backlash. In October and November 2001, boisterous soccer fans, mostly young men, poured into the streets, particularly in Tehran, to celebrate victories and lament losses by Iran's national football team as it struggled to qualify for the World Cup. When young women joined the men in reacting to their team's fluctuating fortunes and some revelers became rioters, the ruling clerics called on security forces to patrol the streets.

The rioters broke windows and shouted antigovernment slogans, most directed at the supreme leader, Ayatollah Khamenei. The security forces responded by arresting some 1,200 people in Tehran alone, most of them under 18. The mounting frustration of youth is eroding not merely the conservatives' credibility, but also the legitimacy of the theocracy as a whole. The soccer spectatorship turned protest provoked [the reformist newspaper] *Nowruz* to sound an editorial warning: "If we are not intent in deceiving ourselves, if we do not wish to reduce the protests of the young generation to the plots of the enemies inspired and guided from the other side of our borders, we must admit that the gulf between the desire for social freedoms and the restrictions and limitations imposed by the government has been turned into the most active form of social conflict. The generation between 15 and 25, which was born and brought up after the Islamic Revolution, is dissatisfied with the limitations and prohibitions that have been imposed on it, not on the basis of the constitution, but rather on the basis of the wishes and demands of a small minority, and is resisting these pressures and compulsions."

# From Revolution to Reform, and Back Again?

The Islamic Republic is a case study of a revolutionary government attempting to reform itself in the face of increased popular appeals for participation by all segments of society in their country's cultural, political, and socioeconomic development. As reformist journalist Harold Reza Jalalipour remarked, "We are witnessing the decline of the fundamentalist movement in Iran. Two decades ago, we had our fundamentalist experience, and we saw the outcome. Fundamentalism is good for protest, good for revolution, and good for war, but not so good for development. No country can organize its society on fundamentalism." Iranians, especially youths, may not wish to forsake religion altogether. Arguably, they hope to see religion returned to the private realm, where one can worship and experience his or her relationship to God without compulsion—not necessarily a post-Islamic Iran, but rather an Iran where citizens may decide freely what role faith plays in their lives.

The future of the reformist movement and its goals, however, remain unclear and even debatable due to the interaction of structural, societal, and human factors. Structurally, the Islamic Republic is, for now, more Islamic than republican, because the constitution enshrines both democratic and theocratic elements, but gives the latter dominance in the management of the state. The exercise of divine sovereignty has so far obstructed the democratic will. Revisions to the constitution are required to reverse this trend, and the conservatives would surely fight this suggestion.

Within society, President Khatami still enjoys support, despite public weariness with the pace of reform. Even after their disappointment with the outcome of the 1999 student revolt, youths returned to Khatami, praising him in heroic terms. Yet young people, including the student movement, lack a coherent leadership and a common strategy. They belong to numerous proreform parties and organizations but have not come together to forge a majority. Youths have engaged the conservatives in a cat-and-mouse game, because they are still not capable of orchestrating a sustained campaign for the cultural and social freedoms they desire. Young Iranians may have to endure more painful confrontations with conservative clerics on the judiciary to gain the experience and will needed for such a campaign.

Nevertheless, youths as well as President Khatami's other constituents have surpassed their president in their conception of reform. In a precarious position with his conservative rivals, Khatami has tried to ensure his own political survival as well as that of the Islamic Republic itself. He is thus more the spokesman for rather than the leader of the reformist movement. A humanist intellectual with a passion for theology and philosophy, he entered the upper echelons of Iran's political arena with some reluctance and little personal ambition. . . .

But as proreform constituents outpace their president in seeking reform, Khatami may eventually have little choice but to rise to the occasion in a moment of crisis and reckoning with the conservatives. If the president does not seize the moment and conservatives continue to resist change, Iranian citizens will become increasingly impatient: their questions already are no longer "Why reform?" or "What kind of reform?" They now urgently ask "How?" and "When?"

# Islam in Iran Is Facing a Crisis

**By Mohammad Khatami**

*Mohammad Khatami was elected president of Iran in 1997 with almost 70 percent of the popular vote. A cleric who played an important role in the 1978–1979 revolution, Khatami campaigned and won on a platform of reform. In September 2002 he introduced two important bills to restructure the government by increasing presidential power and curbing the role of the conservative Guardian Council (a twelve-member body that decides whether laws are compatible with both Islam and the constitution). In this selection Khatami writes that the young Islamic Republic is threatened not only from the outside by Western colonialism, but also from the inside by dogmatic hardliners. Khatami argues that the future of Islam in Iran is dependent on the development of Islamic thinkers who are not only able to avoid being dominated by the West, but are also willing to incorporate the positive features of Western civilization.*

[Iran] confronts a crisis today, and although this crisis is to some extent attributable to global conditions, it is different from the West's crisis. Through our revolution we tried to free ourselves from the shackles of the West's domination. Our revolution made us introspective; we decided to struggle for our independence, to be masters of our own fate. In this regard, we have made great headway in the political, economic, and cultural spheres. But is it possible that we would fall into the West's trap of domination again? This depends on the path we choose in the future and on what the West's own destiny is. The Islamic revolution was a momentous event in the history of the Iranian nation and the Islamic community, and we can rightly say that because of our revolution we have dispensed with many borrowed and Western values that dominated our thinking. Through realizing our own authentic historical and cultural identity, we have laid a completely new

Mohammad Khatami, *Islam, Dialogue, and Civil Society*. Canberra: Australian National University, 2002.

groundwork for regulating our society.

Our revolution proposed the creation of a religiously based system and our society accepted this with enthusiasm and took steps to reach this great goal. The crisis that we experience today can only be remedied if we shed the vestiges of our borrowed identity and don a new garb. Our current crisis is the crisis of birth. . . . Our new civilization is on the verge of emergence. . . .

# Western Dominance

Even though the West has grown old, it maintains tremendous political, economic, military, and technological power, simultaneously wielding a formidable propaganda and communication apparatus to manage the world's perceptions. Equally important, the global economy is controlled and regulated by Western financial institutions.

The West's advanced systems and institutions often legitimize its political power, ensuring its decisive presence in all significant global developments. The military might of Western capitalism is also vast, and even if we concede that official military pacts are not as common as they were, the military and destructive power of the West remains intact.

Politically, the West aims to govern all corners of the world and to dominate the theory and practice of international relations. It possesses the material and symbolic sources of power simultaneously, and it will stop at nothing to achieve its goals and protect its interests. Our struggle with the West is of life-and-death importance.

In its political embodiment, the West does not wish us—or any people—to be independent, free, and masters of our own fate. For if one feature of Western imperialism is violating others' territories and exploiting their economies, the concomitant feature is dominating the world of ideas. The West propagates a world view that lures its prey into subjugation.

We confront a determined enemy that brings all of its material, military, and informational resources to convince us to surrender, or risk being destroyed if we resist. The bitter experience of confrontation between domination-seeking powers and the oppressed masses is too evident to be hidden to anyone.

# The Strategies of Colonialism

In political confrontations the enemy uses the mask of science and culture to deceive us. But in reality its only wish is to induce a people to surrender to its wishes and serve its interests, and to appropriate all of the victims' resources to serve the imperialist power.

Although the West has no qualms about using the most repressive and violent techniques, even its military and overtly oppressive measures are shrouded in seemingly humanistic and misleading guises that divert public opinion from reality.

When colonial powers violate other peoples, they never concede that their aim is to rob the victims' resources or to subjugate them politically. Instead, by abusing their persuasive powers, they try to disguise their crimes through words and ideas that are acceptable to all of humanity. From old times, colonial powers have used the excuse of developing and civilizing other peoples to violate them and rape their lands. Today, like before, the political motto of the West remains defending freedom, human rights, and democracy.

At this juncture our struggle against the West is central to our survival. Any form of reconciliation and appeasement, given the penchant of the opponent for deception, will lead to nothing but our debasement and trampling on our pride. We must struggle against this with all our might, and victory is not beyond our reach. We must depend on God and ask for His guidance, relying on our own historical identity which we have regained through our revolution. With faith in the power of an awakened people and by strengthening the desire for independence and freedom, we must stand firmly opposed to an enemy that lacks humanitarian incentives. This is indeed possible. The awesome resistance of our nation to the conspiracies and crimes of the oppressors can be a lesson for all nations who wish to regain their independence and pride.

# Fanaticism Is Not the Answer

Yet, while ignoring the politically treacherous goals and conspiracies of the West can be catastrophic, we cannot see the West merely in political terms or reduce its whole civilization to political issues. This would also lead us to a harmful dead end.

Western civilization is not limited to its political aspects. Alongside Western politics, there is a system of values and thinking which we must also come to understand and learn to deal with. Here we are confronted with our philosophical and moral opposite, not just with a political rival. To understand the West, the best tool is rationality, not heated, flag-waving emotionalism. Not just here, but nowhere can force offer an effective response to a way of thinking that we consider flawed. That would be self-defeating and counterproductive.

However, mired as they are in shallowness and hype, it is possible that opportunists will take any thought and culture that their audience dislikes and give it political salience and call it a conspiracy to destroy their political base. This does not come from contemplation but from the need to justify their irrational encounter with opposite views, obviating the need for education and a more powerful logic. This is very common among the overly politicized.

Resorting to force is appropriate in confronting a military invasion, conspiracy, or political sabotage. But the way to oppose thought and culture is not through the use of military, security, and judicial means, for using force only adds fuel to the opposite side's fire. We must confront the thought of the opponent by relying on rationality and enlightenment and through offering more powerful and compelling counter arguments. Only comprehensive and attractive thinking can repel this sort of danger. If we do not possess such logic and knowledge, we must endeavor to attain it as our first priority. Islam furnishes us with such a capability. And if some Muslims are devoid of it, the fault lies with them, not with Islam.

If, God forbid, some people want to impose their rigid thinking on Islam and call it God's religion—since they lack the intellectual power to confront the opposite side's thinking on its own terms—they resort to fanaticism. This merely harms Islam, without achieving the aims of those people.

# A Fundamental Difference

In rejecting the West we wish to free ourselves from its political, mental, cultural, and economic domination, for as Muslims, we differ from them fundamentally in world view and values. Thus, to understand our points of contention and to negate the rival's domination, we have no choice but to appraise and understand the West precisely and objectively.

We have to keep in mind that Western civilization rests on the idea of 'liberty' or 'freedom'. These are indeed the most cherished values for humanity in all ages, and to be fair, Western civilization's march from the Middle Ages to modern centuries has broken many superstitions and chains in thought, politics, and society. The West has indeed freed humans from the shackles of many oppressive traditions. It has successfully cast aside the deification of regressive thinking that had been imposed on the masses in the name of religion. It has also broken down subjugation to autocratic rule. These are all positive steps and adaptive to the traditions of creation. Yet, at the same time, the view of the West about humans and freedom has been rigid and one-dimensional, and this continues to take a heavy toll on humanity.

When confronting the opponent in the name of rejecting the West and defending religion, if we stifle freedom we will have caused a great catastrophe. Neither the traditions of creation allow this nor does Islam desire it. But if rejecting the West means critiquing its view of freedom, humanity, and the world, then we will have achieved our most fundamental historical mission.

Indeed, we take issue with the West on the notion of freedom. We do not think that the Western definition of freedom is complete. Nor can the Western view of freedom guarantee human happiness. The West is so self-absorbed in its historical setting and thoughts that it cannot see the calamities that its incorrect view of humanity and freedom has caused. If we look at the West from the outside, we can objectively judge this issue. But achieving this important task requires much intellectual rigor and knowledge. . . .

# Benefiting from the West's Experience

Now, on the basis of our popular revolution we wish to construct an Islamic system. But we can only think of our revolution as giving rise to a new civilization if we have the ability to absorb the positive aspects of Western civilization and the wisdom to recognize the negative aspects of it and to refrain from absorbing them. This means that if we can break through the dead ends that the West has reached because of its values, and pass through them unscathed, we will succeed in our mission.

If we must adopt the positive features of Western civilization, simultaneously casting aside its deficiencies, we have no choice but to understand the West correctly and comprehensively in the first place. We must judge it fairly and objectively and learn from and use its strengths, staying clear of its defects by relying on our revolution's Islamic values. It is obvious that this approach is different from a rigidly political appraisal of the West. Those who cannot separate the political West from the nonpolitical West are acting against the interests of the nation and the Islamic revolution, even though they may be doing so inadvertently. Here, introspection, rationality and objectivity will be effective, not harsh words and violence. . . .

# Religious Dogma

Dogma presents the most formidable obstacle to institutionalizing a system that wishes to provide a model for the present and future of human life, a system based on a more powerful logic than competing schools and ideologies.

The effect of dogma on our society which has a religious identity is vast. Its negative effect is greater than secularism, especially because dogmatic believers usually project the aura of religious legitimacy. Their religious duties compel them to act but they have no connection to authentic Islam, the Islamic revolution, or to the present and the future.

Imam Khomeini[1] especially in the last two years of his life, was deeply concerned with the danger that dogma and backward vision posed to the revolution's path and the progress and welfare of Islamic

---

1. the ayatollah who overthrew Reza Shah Pahlavi's government in 1979

society. In line with all of Imam Khomeini's warnings, vigilance about this phenomenon is crucial to us and the future of the Islamic revolution.

# Religious Intellectualism

Here I want to touch on one of the most important deficiencies of our society at this sensitive juncture, hoping that it spurs debate among thinkers, irrespective of whether they accept my proposition or reject or modify it.

In my view, the greatest defect we have in the sphere of thought and development is the lack or weakness of religious intellectualism, even though I see the ground as ripe for its emergence and growth. . . .

The religious intellectual is one who loves humanity, understands its problems, and feels a responsibility toward its destiny and respects human freedom. She feels that humans have a divine mission and wants freedom for them. Whatever blocks the path to human growth and evolution, she deems as being against freedom.

Our dynamic society at this sensitive juncture badly needs religious intellectuals. If religion and intellectualism are combined, we can hope that our great Islamic revolution will be the harbinger of a new era in human history. But if these two are separated, each will endanger the health of society.

When you mention God to secular intellectuals, they say they prefer to focus on humans. When you mention humans to the dogmatically religious, they say they prefer God. But the religious intellectual seeks 'Godly humans', a creation whose emergence is as pressing a need today as it will always be.

I hope that through our revolution and a well-conceived connection between these two spheres—by connecting religious seminaries and the main centers of thinking in today's world, meaning universities—we will witness the emergence of the religious intellectual. This is a scenario that has neither the deficiencies of secular intellectualism nor those of dogmatic religious belief. Such a movement must shoulder the grand mission of our revolution and solve the crisis that is born out of the birth of a new system, all to benefit humanity, moving us toward a future replete with fulfillment and growth.

# Iran Should Become a Secular Democracy

By Reza Pahlavi

*Reza Pahlavi is the son of Mohammad Reza Pahlavi, the late shah of Iran. Since leaving Iran in 1978 to settle in the United States, Reza Pahlavi (considered by some to be the crown prince of Iran) has been actively campaigning for a secular, democratic government in Iran. In the following speech Pahlavi argues that the exercise of democracy in Iran is not possible under the present theocratic system. In opposition to both Iran's reformist and conservative camps, Pahlavi believes that the separation of religion and state is a necessary precondition for the emergence of democracy, the rule of law, and the enforcement of human rights. He thus favors regime change in Iran, arguing that most Iranians reject the theocracy and want to see the end of political repression and economic decay in their country.*

This evening, I address you as an Iranian citizen committed to a progressive agenda for the future of my homeland, and to the freedoms that my compatriots demand and deserve. This commitment includes the recognition of the important role religion has played historically, and will continue to play in our lives. However, in order to achieve secularism and democracy, I would argue that we must respect and uphold the right of any of our citizens who choose to do so to practice without fear of intimidation or persecution, not only our predominant religion of Shi'ite Islam, but other faiths or systems of belief as well. This must be guaranteed by the future constitution.

To respect religion is not the same as to submit to force, to abdicate one's judgment, or to yield to tyranny disguised as religious mandate. To respect religion in Iran today is to separate it from governance, to assign it the exalted place it deserves in the heart and mind of the individual.

Using Islam to usurp power is to abuse it and ultimately discredit it. This is precisely what the clerical regime has done since its inception. The ruling theocrats have today overwhelmingly lost the trust and support of the Iranian people. In simple terms, religion has been hijacked, by a few, in order to provide a false pretense of legitimacy for a theocratic order that denies the most basic human rights to its citizens.

The regime boasts of the number of presidential and parliamentary elections it has staged in the course of the past 23 years. But cleverly, it omits the glaring fact that elections under its so-called "religious democracy" are limited exclusively to those candidates bearing the seal of approval from the regime. Candidates are only allowed to run on proof of indisputable allegiance to the established leadership. And even when they are elected, their decisions are likely to be reversed by non-elected constitutional bodies. Indeed, such organs as the Guardian Council[1], the judiciary and the office of the faqih (Supreme Leader)— all with their overriding legal powers—are embedded in the constitution precisely in order to override the people's will. The regime, of course, employs various means to induce as many people as it can to participate in its well-orchestrated elections in an attempt to claim legitimacy in the eyes of the Iranian people and the world at large.

# The Impetus for Change

It is now more than five years since the serious inadequacies of the current [2002] theocratic regime in Iran, evident to the majority of Iranians, have also come to the attention of the international community. This awareness has come about mostly as a result of the re-emergence of Iranian youth on the political scene. This is to be expected. Nearly 50 million of Iran's 70 million citizens are under the age of 30. These young people desperately need and demand freedom, jobs, housing, education, healthcare, and economic opportunity. They hold the key to Iran's transition from religious totalitarianism to a secular representative government, complete with economic promise, a civil society and guarantees for liberty, gender equality and a better life.

The impetus for change in Iran's political environment is to a great extent a consequence of this resurgence. Our youth are the vanguard of the movement for change and have achieved considerable success in undermining the hardliners of the theocratic government. The student rebellion initially met brutal suppression in July 1999. However,

---

1. a board of twelve jurists with the authority to veto laws passed by parliament if they are found incompatible with Islamic law

neither imprisonment and torture, nor various intimidation tactics perpetrated by their rulers, discouraged them from continuing the struggle for liberty. Today, this struggle is, in fact, spearheading a national crusade against theocratic rule and is redefining the very role of religion in our society.

What our youth demand is what has been historically sought by their counterparts in free societies the world over. They no longer accept the suffocating space and the sterile intellectual atmosphere ordained for them by their rulers, whom they consider abusers of religion in pursuit of unholy agendas.

Particularly noteworthy is the valiant role Iranian women have played in defying the clerical establishment. Constituting 51 percent of the population, Iranian women were the first to bear the brunt of the regime's suppression. They were among the first to rise against the tyranny of a system that from its inception sought to force them into the confines of a second-class citizenry. On this defining issue, it is clear that the regime's inherent failure lies in its dogmatic rejection of equal rights for women. Similarly, its denial of equality under the law for religious and ethnic minorities is yet another glaring violation of the fundamental principles of human rights.

# The Law of the People

The failure of the theocratic system to resolve Iran's serious socioeconomic problems has caused a growing number of Islamic theologians, who themselves were founders and theoreticians of the Islamic Republic, to openly question the very doctrine of "velayat-e faqih" [Supreme Leader], although many still promote the contradictory concept of "religious democracy."

More importantly, the people today attribute these shortcomings to the root cause: the clash between theocracy on the one hand, and modernity and democracy on the other. Iranians today clearly understand and openly debate this principal: that democracy presupposes the sovereignty and inalienable rights of the individual in the context, not of divine law, but of the law of the people.

Democracy is based on the free expression of thought and respect for human rights, including full recognition of equal rights for women and for ethnic and religious minorities. A system such as the Islamic Republic, in which the Sovereignty of God exercised through the faqih and his paraphernalia of governance is intricately woven into the constitution, can never become democratic.

Since the abrogation of its Constitution would amount to loss of raison d'être for the regime, and thus would never be volunteered by the ruling clergy, only a complete "regime change" could usher in real democratization. It is clear that other than promoting an illusion of democracy and thus prolonging the era of political repression and economic decay in Iran, these "reformed theologians" fail to provide effective solutions for rectifying popular grievance, reviving Iran's economy, and rebuilding the country's damaged relations with the outside world.

The rift between the regime and the people is widening daily. The regime is losing legitimacy in the eyes of the people; but it must persevere in its ways in order to maintain legitimacy in its own eyes. People have lost, and are increasingly losing, their cherished beliefs in Islam because religious institutions and practices are inextricably intertwined with the failed institutions of the government. They are also confronted with a dilemma of colossal proportions in that they are faced with a judicial ruse. On the one hand, the judiciary claims independence, which is how a good judicial system ought to be; on the other hand, the judicial system is constitutionally and therefore practically biased against the fundamental rights of the people, which is precisely what a good judiciary ought not to be. As such, the real struggle today is between the theocracy, and the people who pursue modernity, secularism and democracy.

The clerical regime is both unwilling and unable to deliver the types of reforms that can begin to address people's fundamental needs. After 23 years of despotism and sharp socio-economic decline, most Iranians reject the current regime and more than ever wish to free themselves from the shackles of a medieval system, clearly out of tune with the needs of a modern society. The regime's efforts to curb dissent and ignore the public's outcry has proven ineffective in preventing the Iranian people's march towards a secular and progressive society—one in which state and religion are once and for all separated. In essence, the people of Iran have reached the conclusion that the system is inherently non-reformable, and that theocracy and democracy are incompatible.

# Islamic Terrorism

Let me turn to the issue of terrorism and the Islamic regime in Iran. The clerical rulers of Tehran cannot become loyal partners in the global war against terror. In its 23 years of rule, the Iranian theocracy has in fact used terror as an instrument of policy. The prime victims of this practice have of course been the people of Iran. But the regime has also championed terrorism of global reach, and since 1983 persistently topped the lists of states sponsoring terrorism. The record is unmistakable. Details are set forth in official reports of the United Nations, Amnesty International, the U.S. State Department and numerous other sources. More tangibly, they are reflected in terrorist indictments against the most senior Iranian officials, issued by courts in Germany and the United States.

Let there be no doubt, similar to the old Soviet doctrine of "communism international," the clerical regime's raison d'être is the export of the "Islamic revolution," first regionally and then globally. This is embedded in the very same constitution that the present leaders have sworn to uphold at any cost.

No wonder the involvement of the Islamic regime in terrorist activities stretches well beyond the Persian Gulf, to Europe, Africa and Latin America. Furthermore, having lost legitimacy domestically, the regime is in dire need to score points beyond its borders in order to retain such legitimacy in the eyes of all extremists, from [Osama] Bin-Laden to others. So long as the Islamic regime in Iran exists as a model and epicenter, it would provide solace to radical Islamists across the world, and as we have realized, such a role is far more dangerous and pernicious than weapons of mass destruction.

As a pivotal country in the Middle East and the Persian Gulf region, and with the largest population and one of the oldest civilizations, I am convinced that the institutionalization of democracy, secularism and the rule of law in Iran will have positive ramifications, not only for our country, but also for our neighbors. A secular and democratic Iran will be a force for stability and moderation in that volatile region and consequently a positive and constructive influence for the promotion of international peace and security.

# Two Views of Iran

Finally, let me address a few points regarding US Foreign Policy and the international reaction and posture vis-à-vis Iran. There are two categories of countries or governments: those who separate the people of Iran from their rulers, and those who still believe the conflict to be one between two camps, the so-called moderate and conservatives. It appears that the current US administration has finally shifted to the first group. Subsequently, the symbolic gesture from Iranian citizens, holding a candle light vigil subsequent to September 11th, was acknowledged and responded to by the President and his administration. For the most part, these gestures and demarcations were positively interpreted and received by most Iranians, to the detriment of their disagreeing rulers.

The European Union, on the other hand, appears to still be stuck with the old cliché, and has succumbed to a carefully orchestrated good cop/bad cop game masterfully played by Tehran. What preoccupy most Iranians—myself included—are the ongoing negotiations between EU representatives and the clerical regime. It is imperative that any trade considerations should be preceded by major changes in the regime's domestic behavior in the overall context of human rights violations. The worst thing that could happen is for the world to condone these violations while pursuing short-term economic interests, and to be literally throwing a lifeline to a sinking regime. That will not bode well for people's morale, and will in fact alienate them vis-à-vis all those who chose to ignore their plight at this critical juncture. I therefore caution the world community in realizing the consequences of their actions and policies regarding Iran, particularly in the short term, and in light of recent dramatic upheavals.

Having said that, I have told my compatriots time and again, that we should not depend on anybody but ourselves in our nonviolent fight for freedom, democracy, and progress. We do not expect other nations to have our interest at heart more than their own. We expect them, however, to recognize that a civil and reliable government in a country like Iran, in a region like the Middle East, is to everyone's interest. And for advocates of freedom and human rights, we hope that they will continue to stay true to their stated principles, especially when they clearly witness the plight of our people under the rule of the Islamic Republic.

# What the World Can Do

But let me emphasize this: There is no "single formula" for the Middle East. Iran need not be confronted with military action. Iran's problem will be resolved by Iranians alone. Naturally, in bringing momentum and direction to the process of change, we expect the world to give moral support to our people, thus further empowering acts of civil disobedience and the quest for liberty and secularism.

Unlike the 20th century when governments invested in regimes, the 21st century will prove that ultimately investment in people and democracy far outlasts investment in unpopular regimes.

Our world has witnessed the dawn of new democracies brought about by nonviolent civil disobedience movements, from Africa to Latin America and throughout Eastern Europe. Let there be no doubt that Iranians thirst for the same chance to restore their inalienable right to self-determination, thus restoring the civility, dignity, tolerance and sovereignty for which my homeland was known for so many centuries.

The world must care and make the right choice in favoring the winds of change that will usher in secularism, human rights, and democracy for Iran—reversing the cycle of violence, and directly translating into regional peace and stability for the world.

**CHAPTER 3**

# Iran's Foreign Relations

# Iran Poses a Nuclear Threat

**By Jon B. Wolfsthal**

*Iran is party to the Nuclear Nonproliferation Treaty, which allows member states to acquire the means to produce nuclear power for peaceful uses. However, many suspect that although Iran claims that it wants to produce nuclear power for fuel, it really seeks to develop full-scale nuclear weapons. In the next article, Jon B. Wolfsthal argues that a nuclear-armed Iran poses a threat to regional and global security and would encourage an arms race in the Middle East. In order to discourage the proliferation of nuclear weapons in Iran, the United States and Russia must adopt a new strategy, including possibly guaranteeing a foreign supply of nuclear energy to Iran. Wolfsthal is the deputy director of the Nonproliferation Project at the Carnegie Endowment for International Peace.*

Even during the depths of the Cold War, the United States and the Soviet Union often worked together to halt the spread of nuclear weapons to new countries. Now, both countries are dealing with the realization that Iran's nuclear program is more advanced than previously thought and may be aimed directly at acquiring nuclear weapons in the next few years. Unfortunately, the approaches being pursued by both countries will do nothing to slow Iran's ability to produce nuclear weapons, and a new approach and better coordination is desperately needed before it is too late.

## Flawed Approaches

For the better part of a decade, U.S. officials pressured Russia to stop its support for the Bushehr nuclear reactor project in Iran. The United States argued that the power plant was a front for Iran to acquire weapons-related technology, a charge that Russia rejected. It now appears that both sides may have been wrong.

Counter to U.S. projections, Iran appears to have used Pakistan and other third parties to develop a uranium enrichment technology based on centrifuges, instead of relying on covert acquisitions of Russian technology. This does not mean, however, that Russian experts or companies have not been involved in this program without the Kremlin's knowledge or permission—only that Russia appears not to be the primary source of Iran's newfound capabilities. Yet Russia also ignored clear signs that Iran was interested in much more than a peaceful nuclear power program. Its willingness to engage in nuclear commerce with Iran, while financially beneficial, is now coming back to negatively affect Russia's security.

To remedy the situation, the two countries have adopted similarly flawed approaches. Russian officials are working with Iran to ensure that any fuel used in the reactor at Bushehr—fuel that when reprocessed could produce hundreds of nuclear weapons worth of plutonium—is returned to Russia. For its part, with Russian support, the United States is pushing Iran to join the International Atomic Energy Agency's enhanced inspection agreement, which will give the agency broader inspection and monitoring rights in Iran.

# The Nuclear Nonproliferation Treaty

While both of these initiatives are helpful, they will do absolutely nothing to head off the main challenge posed by Iran's growing nuclear program—Tehran's construction of advanced centrifuge enrichment facilities that could produce enough weapons-grade uranium for 20 weapons per year by the end of the decade. Iran has stated that it is developing the means to produce its own enriched uranium fuel for the Bushehr reactors out of concern that the United States will convince Russia to cut off its fuel supply.

Under the Nuclear Nonproliferation Treaty [NPT], to which Iran is a party, states are entitled to engage in all manner of peaceful nuclear development as long as they accept international inspections. This provision, however, allows states to use the cover of the treaty to acquire the very means to produce a formidable nuclear arsenal, and then later withdraw from the pact and use the material for nuclear weapons. At the heart of international concerns is the risk that Iran will follow just this scenario to the detriment of regional and even global security.

To head off this eventuality, the United States and Russia should reach quick agreement on a new strategy that would not only head off Iran's nuclear weapons potential, but address the underlying flaw in the NPT system. At a minimum, Russia should offer to guarantee— with explicit U.S. endorsement—Iran's supply of fuel for the Bushehr reactor as long as Iran abandons its indigenous uranium enrichment and plutonium production programs. This offer would give Iran a clear choice—a reliable foreign source of nuclear energy or an internal nuclear program with weapons potential. The choice that Iran makes would help show the international community Iran's true intentions.

# We Must Deter Nuclear Competition

To many, it is already clear that at a minimum, Iran is seeking the option of producing nuclear weapons through its own independent nuclear program. Given its history of conflict with Iraq—a state by no means guaranteed of a peaceful and stable future—as well as the perceived threats from Israel's and America's nuclear arsenals, Iran's position is understandable in some circles. But this nuclear option would only serve to increase the desire of other countries, including Saudi Arabia, Syria and even a future independent Iraq, to acquire their own nuclear options, to say nothing of the steps Israel might take before Iran's became a reality.

Thus, in addition to the offer to guarantee Iran's supply of low enriched uranium fuel for its nuclear reactor, the United States and Russia should revisit the idea of establishing a clear policy that nuclear weapons will not be used to threaten states that do not have nuclear weapons or an active nuclear program. Amazingly, since the end of the Cold War, both the United States and Russia have increased the circumstances under which they would be willing to use or threaten use of nuclear weapons. It is time the two countries recognize that such a policy has negative implications that could drive states to acquire nuclear weapons.

Russia and America have an important legacy of preventing proliferation of which they should be proud. It is a legacy that should be revived and focused on the core proliferation threats in Iran and elsewhere before the nuclear confrontation of the Cold War is replaced by a broader nuclear competition the two states will not find as easy to control.

# Iran Sponsors Terrorism in the Middle East

**By Michael Rubin**

*On January 3, 2002, the Israeli navy intercepted the* Karine-A, *a freighter owned by the Palestinian Authority, and discovered fifty tons of sophisticated Iranian weaponry bound for Gaza. The $15 million cargo included 346 rockets, 700 mortar bombs, several hundred rocket-propelled grenades, and two tons of C-4 plastic explosives. Because of the Palestinian Authority's strong links to anti-Israeli terrorist groups, the weapons posed a serious threat to Middle East peace. In the following selection, Michael Rubin writes that the Iranian sale of weapons to the Palestinian Authority is but one example in a long history of Iranian support for terrorism. Despite reformist rhetoric, the current regime is strategically engaging with a number of terrorist groups in order to advance its own interests, and poses a threat to all of the Middle East. Rubin serves as an Iran and Iraq policy adviser at the Pentagon. In 2001 he published* Into the Shadows: Radical Vigilantes in Khatami's Iran.

W hile the Iranian government has always maintained close links with Hezbollah and other Islamist groups, the Islamic Republic has long had a strained relationship with [Palestinian leader Yasir] Arafat. Immediately after the Islamic Revolution, Arafat antagonized his Iranian patrons when he threw his support behind [Iraqi dictator] Saddam Hussein in the Iran-Iraq War. However, the *Karine-A* affairs shows incontrovertibly that Arafat and the Islamic Republic have now put aside their differences and are cooperating to advance Iran's vision of the Middle East. The direct sale of such sophisticated weaponry marks a clear strategic escalation in the region.

Michael Rubin, "Iran and the Palestinian War Against Israel: Implications of the *Karine-A* Affair," *The Middle East Backgrounder*, February 2002. Copyright © 2002 by the American Jewish Committee. Reproduced by permission.

# Iran Is a Threat to Its Neighbors

The Islamic Republic of Iran unapologetically sponsors terror. On January 31, 2002, Supreme Leader Ayatollah Ali Khamenei defended groups responsible for the deaths of hundreds of Americans. Speaking in Tehran, he declared, "The only sin of [terrorist organizations] Hamas, Islamic Jihad, and Hezbollah of Lebanon . . . is that they have taken practical action. . . ." If terrorism is merely practical action, then the Islamic Republic has a long and deadly history of being practical.

Just shy of one month after the World Trade Center and Pentagon attacks, President Bush announced the creation of a list of the 22 most-wanted international terrorists. No Iranian officials made the list, but almost one-third of the most-wanted received direct support or safe-haven from the Islamic Republic. Four of the terrorists are suspects in the 1996 bombing of a U.S. Air Force housing facility, Khubar Towers, near Dhahran, Saudi Arabia, which killed 19 Americans and wounded more than 500 others; three others receiving safe-haven or Iranian support were members of Lebanon's Hezbollah.

Despite Iranian officials' recent reformist rhetoric, such direct association with terrorists should come as no surprise. The U.S. State Department's most recent report on *Patterns of Global Terrorism* labeled [President Mohammad] Khatami's Iran "the most active state sponsor of terrorism in 2000." Iran actively supported terrorist groups such as Hezbollah, Hamas, Palestinian Islamic Jihad, and Ahmad Jibril's Popular Front for the Liberation of Palestine-General Command [PFLP-GC]. Just three weeks before the World Trade Center attacks, the latter organization called on "the Arab and Islamic Nation to strike all American and Zionist interests. . . ."

Beyond the view of the Western press, Iran continues to support terrorism in neighboring countries, such as in secular and pro-Western Turkey, as well as the U.S.-protected Kurdish safe-haven of Iraq. Iranian intelligence has significantly increased its presence in Afghanistan. Despite promises to the contrary, Iran has never revoked the *fatwa* [legal decree] calling for the murder of British author Salman Rushdie.

# Iran Supports Terrorism

Even if no Iranian citizen made the most-wanted terrorist list, one Iranian official has been directly implicated in anti-American terror: Ahmad Sharifi, a brigadier-general in Iran's elite Islamic Revolutionary Guard Corps is widely believed to have coordinated the 1996 Khubar Towers bombing [in Saudi Arabia]. In 1999, State Department spokesman Jamie Rubin revealed that Washington had "specific information with respect to the involvement of Iranian government officials." In

August 1999, Bill Clinton wrote to Khatami seeking the Iranian president's assistance in investigating Sharifi. Khatami flatly refused to cooperate in any way in the Khubar Tower bombing, even if such counterterror cooperation would improve relations.

Iranian involvement in terrorist attacks is not the work of isolated vigilantes. Many U.S. government officials dispute that any serious division exists between top Iranian officials and the vigilantes. Then-Director of Central Intelligence James Woolsey in 1994 declared that Iranian terrorist attacks are "not acts of rogue elements," but rather are "authorized at the highest levels of the Iranian regime." European investigators agree. In 1997, after an almost four-year trial, a Berlin court found a committee consisting of no less than Supreme Leader Khamenei himself, the president, the intelligence minister, and the foreign minister to have ordered assassinations of Iranian dissidents in a Berlin cafe. Amnesty International declared that the verdict "provides further evidence of [an] Iranian policy of unlawful state killings."

Iran has made little secret of its investment in terror. Patrick Clawson, a former World Bank economist who is currently research director at the Washington Institute for Near East Policy, investigated the Islamic Republic's budget and found that Iran allocated approximately $75 million annually for terrorist activities. Following the death of his daughter at the hands of a Palestinian Islamic Jihad suicide bomber, Stephen Flatow sued the government of Iran for funding the group. In his findings of fact, United States District Court Judge Royce C. Lamberth noted that "the Islamic Republic of Iran is so brazen in its sponsorship of terrorist activities that it carries a line item in its national budget for this purpose." With the Islamic Republic willing to put substance behind its threats, Iran's rabidly anti-Israel rhetoric cannot be dismissed.

A U.S. strategy must be broader than a renewed focus on Iran. Several countries and groups actively enable Iran's sponsorship of terror and violent opposition to regional peace and security.

By virtue of its geographical location, as well as through its military and political domination of Lebanon, Syria provides the means to enable the Iranian leadership to provide direct support to Hezbollah, Palestinian Islamic Jihad, and other terrorist groups. Even as Syria was elected to the United Nations Security Council on October 8, 2001, the Syrian government continued to host more international terrorist groups than any other nation. Hamas opened a new main office in Damascus in March 2000. Also maintaining headquarters in Damascus are the Popular Front for the Liberation of Palestine-General Command, Palestine Islamic Jihad [PIJ], and George Habash's Popular Front for the Liberation of Palestine [PFLP]. Syria grants these groups and others permission to operate camps or safe-houses in

Lebanon's Bekaa Valley, which remains under Syrian occupation. . . .

The "Party of God" [Hezbollah] has a long relationship with Iran and depends upon the Islamic Republic for material support, expertise and training, and safe-haven. While Hezbollah claims that it seeks only to "resist the occupation of Lebanese national soil," its goals are much greater. By colluding in the attempted import into the Gaza Strip and West Bank of missiles capable of bringing down civilian jetliners or striking at the heart of Israeli cities, Hezbollah has demonstrated that it is an unreformed terrorist group undeserving of the political legitimacy ascribed to it by many European Union and other diplomats. The only way to prevent wider war in the Middle East is to reign in terrorism. And terrorism must remain a black-and-white issue. To engage in any way with Hezbollah would be to justify some forms of terrorism as somehow acceptable.

# Who Is Iran Sheltering?

While hosting Ahmad Sharifi is indictment enough of the Islamic Republic's support for terror, the Iranian government continues to shelter individuals far more deadly. Imad Mughniyeh is a case in point. Born in Lebanon in 1962, Mughniyeh was a founding member of Hezbollah and director of the group's overseas operations. His trail of blood clearly shows that Hezbollah is not merely some resistance army engaged in a struggle against Israel, as many European diplomats assert. The following are just some of the acts traced to Mughniyeh who has, since 1991, called Iran home:

The remote-controlled truck bombing of the U.S. Embassy in Beirut on April 18, 1983, killing 63 employees and wounding 120 others.

The suicide truck bombing of the U.S. Marine headquarters in Beirut on October 23, 1983, killing 241 soldiers and wounding 81. The Marines were in Beirut as part of a peacekeeping mission. By targeting the peacekeepers, Mughniyeh fulfilled the Iranian objective of prolonging the Lebanese Civil War.

The January 18, 1984, murder of Malcolm Kerr, president of the American University of Beirut [AUB].

Throughout the 1980s, planning the kidnappings of a number of Western citizens for use as hostages, including CNN's Beirut bureau chief Jeremy Levin; the Associated Press' Terry Anderson; the Reverend Benjamin Weir; Church of England mediator Terry Waite, and AUB officials Frank Reed, David Jacobsen, and Joseph Cicippio, among several others.

Personally torturing and murdering hostage William Buckley, the CIA station chief in Beirut, who was kidnapped on March 16, 1984, and whose body was recovered on December 27, 1991.

On June 14, 1985, Mughniyeh was among the Hezbollah members who hijacked TWA flight 847 from Athens to Rome with 145 passengers and eight crewmembers on board. During the course of the 17-day crisis, the hijackers murdered one American passenger. Off-loaded in Beirut, the hijackers separated the passengers with Jewish surnames.

More recently, Mughniyeh is alleged to have planned the 1992 bombing of the Israeli embassy in Buenos Aires, and the destruction of that city's Jewish cultural and communal center two years later.

In addition to his deadly past, Israeli intelligence officials believe that Mughniyeh may have sought to promote future bloodshed by serving as the middleman for the *Karine-A* arms shipment.

# Iran's Deadly Protégé

While Mughniyeh and Hezbollah may comprise one of the deadliest terrorist groups in the Middle East, Iran sponsors numerous other Islamist terror groups capable of wreaking just as much death and destruction to support the Islamic Republic's political aims. Perhaps the group with the closest ties to the Islamic Republic is Palestinian Islamic Jihad [PIJ]. PIJ became active in the West Bank and Gaza in 1979 and seeks to eradicate Israel. With offices in Beirut, Damascus, Khartoum, and Tehran, PIJ has always looked at Iran as its chief ideological patron. According to Reuven Paz, academic director of the International Policy Institute for Counter-Terrorism at the Inter-Disciplinary Center in Israel, PIJ is "the only group in the entire Sunni Arab world that wholeheartedly supports the Iranian Islamic revolution and the Iranian regime." Indeed, founders Fathi Shqaqi, 'Abd al-'Aziz 'Odah, and Bashir Musa embraced Ayatollah Khomeini's concept of *vilayat-i faqih* [Guardianship of the Jurisprudent], the theological underpinning of religious dictatorship. Consequently, while PLO chairman Arafat long had strained relations with the Islamic Republic (due to Arafat's consistent support for Iraqi leader Saddam Hussein), the PIJ has remained close to Khamenei, Khatami, and the spectrum of the Islamic Republic's leadership.

Palestinian Islamic Jihad was behind a rash of suicide attacks in the mid-1990s that derailed the peace process and contributed to Benjamin Netanyahu's victory over Shimon Peres in Israel's 1996 elections. Specifically, Islamic Jihad claimed responsibility for a January 22, 1995, bomb attack near Netanya that killed 18 Israelis, as well as an April 9, 1995, attack on a civilian bus that killed American Alisa Flatow, among several others. On March 4, 1996, a PIJ suicide bomber detonated a 20-kilogram nail bomb in a Tel Aviv shopping mall slaughtering 20 civilians, mostly teenagers and senior citizens.

Despite giving safe-haven to Mughniyeh and underwriting the activities of Palestinian Islamic Jihad, President Khatami declared in a November 9, 2001, *New York Times* interview that "there are no terrorists in Iran." Such statements by Khatami should not be taken at face value. Ever since his triumph in the 1997 presidential elections, Western diplomats, journalists, and policymakers have favored a non-confrontational approach and have been willing to overlook such inconveniences as Iran's support for terrorism. Believing that Khatami was sincere in his rhetorical outreach to the West, the logic went that to hold Iran accountable for her actions would damage the cause of reform. For example, prior to British foreign secretary Jack Straw's September 2001 visit to Tehran, one British official commented, "There is no reason the foreign secretary's visit would be soured by raising the issue of this man [Mughniyeh].". . .

# A Conference for Terror

The real goal of Iranian government policy was prominently displayed just a week after the capture of the *Karine-A*. Beginning on January 9 [2001], leading officials of Hezbollah, Hamas, Palestine Islamic Jihad, as well as prominent personalities from Iran, Syria, Lebanon, and the Palestinian Authority gathered in Beirut for a two-day conference. Iranian parliamentarian 'Ali Akbar Mohtashemi, an adviser of Khatami and publisher of a banned reformist paper, brought the endorsement of the Iranian government.

Ayatollah Sayyid Muhammad Hussein Fadlallah, spiritual leader of Hezbollah, gave the keynote speech, sermonizing that all Muslims should promote the cause of "martyrdom operations," a euphemism for suicide bombings. Sheikh Hasan Nasrallah, political leader of Hezbollah, castigated respected Muslim clerics who had condemned suicide bombings, calling such religious prohibitions "fatwas to surrender." The conference concluded that such "martyrdom operations are national principles that cannot be changed."

The Middle East is already seeing the fruits of the Iranian government's recent acceleration of terror sponsorship. At the conclusion of the conference, Hamas official Musa Abu Marzook declared in an interview with CBS's *60 Minutes* that Hamas was developing a missile with a range long enough to hit most Israeli cities. The Kassam-1 rocket has a range of about three miles, while the Kassam-2 rocket, currently in development, will have a range double that, as well as an increased payload. Abu Marzook's statements were not empty rhetoric. On February 10 [2002], Palestinians fired Kassam rockets from the Gaza Strip into Israel. Iranian-sponsored coordination of secular, nationalist, and Islamist terror groups is no longer merely a threat; it has become reality.

# Iran Does Not Support International Terrorism

By Kamal Kharrazi

*From 1980 to 1989 Kamal Kharrazi was the president of the Islamic Republic News Agency. He went on to represent Iran at the United Nations and has served as Iran's foreign minister since 1997. In the following speech, Kharrazi reaffirms Iran's support for both intifada, the Palestinian revolt against Israeli occupation in the Gaza Strip and the West Bank, and the UN fight against international terrorism. He argues that intifada is a legitimate form of defense, and that Iran's support is in compliance with international norms. According to Kharrazi, the policies of the United States and Israel actually promote terrorism by creating an insecure climate in the Middle East. Furthermore, he states that the United States and Israel accuse Islamic states and Palestinian groups of terrorism as a justification for their own aggressive military actions against these states and groups.*

In the name of the people and the Government of the Islamic Republic of Iran, I reaffirm our full support for Intifada, condemn, in the strongest possible terms, the Israeli savage aggression against Palestinian people and their leaders, and call for coordination and unity of the Muslim Umma [community] to help counter these crimes.

## We Must Define Terrorism

Terrorism is a major challenge to international peace and security. The September 11th attacks in the United States amply demonstrate the

Kamal Kharrazi, statement before the Extraordinary Session of the Islamic Conference of Foreign Ministers on Terrorism, April 2002.

fact that terrorism recognizes no boundary between the rich and the poor, or between the innocent and the guilty. It threatens anybody or any entity in the world regardless of geography, status or power.

Therefore, fighting terrorism requires an internationally recognized legal framework that is supported by concerted efforts and political resolve of all nations and key players in the international community. In this context, all countries should organize their efforts under the auspices of the United Nations to adopt a comprehensive counter terrorism strategy and develop a clear definition of terrorism which embodies all forms and manifestations of terrorist acts. Under such circumstances, the fight against terrorism would become more effective, transparent and law-based, precluding double standard practices or selective attitudes on the part of the world powers. In this process the distinction between terrorism and legitimate defense against foreign occupation or military aggressions must be defined. . . .

The campaign against terrorism has always been a top priority for the Islamic Republic of Iran that has been a victim of this utterly appalling act. In our opinion, protecting our national security and preserving the safety and security of our citizens against terrorism compel us to develop a global view of terrorism. In his message of September 16th, 2001, addressed to the UN Secretary General, the President of the Islamic Republic of Iran called for a world summit on eradication of terrorism. During the 56th Session of the United Nations General Assembly [in November 2001], President Khatami also stressed the imperative of developing a "coalition for peace" in order to embark on a serious and effective way to muster collective resolve and responsibility to fight terrorism and promote international peace and stability. . . .

# The Aftermath of September 11th

The September 11th tragedy and its dire consequences for the international community ought to be viewed from various angles. In the aftermath of the attacks on America, combating terrorism became a top priority for the entire international community and for the United Nations. Great expectations were raised for adoption of an effective global plan of action against terrorism based on wisdom and rationality and in accordance with the principles embedded in international law. To the regret of many, certain major powers, especially the United States, have severely damaged this constructive international momentum against terrorism through a series of unilateral actions, shortsighted policies and arrogant statements. Repeated provocative statements by senior US officials, particularly the State of the Union Address by the President of the United States on 29 January, 2002, were filled with heavy-handed observations, baseless accusations and threats against certain countries as well as some Islamic movements and freedom fighters in Palestine and Lebanon. They clearly point to a hegemonic policy that runs counter to the present realities of the world, and unfortunately serve to sustain terrorism by acting as a divisive factor in the world resources and potential that could otherwise be mobilized against terrorism. Unfounded and biased claims against other states, groups or individuals and branding them as the perpetrators or supporters of acts of terrorism will undoubtedly leave behind far-reaching negative impacts on the confidence and trusts, which is a key ingredient for promoting international peace and security. The Western media has also concurrently launched a large scale campaign associating Islam and Muslims with terrorism. The result has been a mounting psychological, social and political pressure against Muslims, especially Muslim communities in the West, some of whom have been physically hurt, their civil rights as individuals severely compromised and most of them fear for their lives and security. . . .

# Israel and Intifadah

The Zionist entity [Israel] made the most sinister use of the anti terrorism climate created by the September 11 attacks. It unleashed its military and security forces, armed to the teeth with the most sophisticated American weapons, against Palestinian children armed with stones, but bravery matched by none. Merciless killing of hundreds of defenseless Palestinian civilians, deliberate bombardments of residential homes, office buildings, and hospitals, assassination of Palestinian leaders, and imposition of blockades on Palestinian areas are all clear examples of Israeli state terrorism which many in a position of power conveniently choose not to see.

Intifadah is the legitimate response of the Palestinian people on the basis of their inalienable rights of self defense and self determination and in compliance with all international norms to counter foreign occupation. . . .

It is an imperative for Islamic nations to assess the current developments and its dimensions in order to mobilize their resources and capacities in support of practical and strategic decisions in dealing with the international terrorism, and taking concrete steps to support the Intifadah of the Palestinian people to counter the Israeli policy of oppression. We can achieve these important goals only through enhanced coordination and cohesion among Islamic countries.

The Islamic countries must counteract any political move or campaign intended to identify Muslims or Islam with terrorism and violence. We should coordinate our efforts to confront the Israeli and others' attempts and diplomatic maneuvers to level baseless allegations of terrorism against Islamic states and Palestinian groups including associating them with the lists of those suspected of involvement in terrorism. . . .

In conclusion, I wish to reiterate once more that the goal of eradicating terrorism could only be realized through the participation of all members of the international community on the basis of the principles enshrined in international law and in the UN Charter.

# The United States Must Support Iran's Reformers but Counter the Ruling Regime

By Richard L. Armitage

*Richard L. Armitage is the U.S. deputy secretary of state. From 1983 to 1989 he served as assistant secretary of defense for International Security Affairs at the Pentagon, where he played a leading role in developing Middle East security policies. In the following testimony given before the Senate Foreign Relations Committee in 2003, Armitage contends that the United States must continue a dual policy toward Iran that both counters the regime's various threats to international security and supports the efforts of the country's reform movement. The United States is most concerned with the ruling regime's abuse of human rights, its nuclear weapons program as well as its chemical and biological weapons program, its support for terrorism, and its interference in regional politics. At the same time, the United States seeks to encourage the Iranian democratic reform movement by providing educational support.*

I ran is a country in the midst of a tremendous transformation, and I believe American policy can affect the direction Iran will take. This is a complex situation, but if you will allow a simplification: today in Iran, there is a struggle between destructive elements of Iran's society and leadership, who want to keep the country mired in a violent, cor-

Richard L. Armitage, testimony before the Senate Foreign Relations Committee, October 28, 2003.

rupt, and insular past, and a forward-looking popular movement, which wants a more engaged and modern Iran to emerge. The fact that the Nobel Peace Prize was just awarded to an Iranian citizen [Shirin Ebadi] is no aberration; rather it is a sign of the sweeping desire for change across Iranian society. Indeed, all Iranians stand to benefit from a modern state, one that draws on the strengths of free minds and free markets. American and international security and well being also stand to benefit. United States policy is, therefore, to support the Iranian people in their aspirations for a democratic, prosperous country that is a trusted member of the international community.

Given the complexities of the situation, it is no surprise that there is a range of views—including on this [Senate Foreign Relations] Committee—about how to best implement that policy. That is entirely appropriate. Indeed, a single, static, one-size-fits-all policy would not be appropriate in the circumstances. In order to best protect and advance U.S. interests, our policy needs to be flexible, dynamic, and multifaceted. That is why the President and this Administration are pursuing a policy that weighs the full range of options available to us, both through bilateral and multilateral means. We seek to counter the government of Iran's negative and destructive policies and actions, while encouraging constructive policies and actions and engaging in a direct dialogue with the Iranian people about the freedoms they want for their own country.

As President Bush noted when talking about Iran [in October 2003] . . . , not every policy issue needs to be dealt with by force. Secretary [Colin] Powell also noted that we do not seek conflict with Iran. We will continue to pursue nonproliferation and other such control measures as necessary and we must keep all available options on the table, given the lack of clarity about Iran's future direction and ultimate destination. At the same time, we are prepared to engage in limited discussions with the government of Iran about areas of mutual interest, as appropriate. We have not, however, entered into any broad dialogue with the aim of normalizing relations.

# The Challenges Posed by Iran

There is no question that Iran is engaged in a number of destructive policies and actions. Our most pressing concerns are Iran's poor human rights record, nuclear weapons program, as well as chemical and biological weapons programs, support for terrorism, and interference in regional politics, particularly in the Arab-Israeli peace process. These behaviors, along with the government's oppressive and corrupt centralized economic policy, shake the confidence of the international community and deny the Iranian people the quality of life commen-

surate with the country's rich human and natural resources. These behaviors also undermine regional stability and have ripple effects across
U.S. and international security. We are taking and will take the necessary measures to protect U.S. interests.

Across the board, the United States is actively countering such Iranian activities through a variety of tools, including sanctions, interdic-
tion, law enforcement, diplomacy, and international public opinion.
When necessary, we will act alone. The United States, for example,
has a broad array of sanctions on Iran. This includes prohibitions on
a range of exports and assistance, particularly to the military and to
the oil industry, strict regulations on economic transactions, and targeted sanctions against specific entities in other countries that aid
Iran's weapons of mass destruction programs.

We believe, however, that international and multilateral responses—
if sustained—will be especially effective in meeting the challenges
Iran poses to regional stability, disarmament and nonproliferation
regimes, and the rights of its own citizens. As President Bush said, we
have confidence in the power of patience and the collective voice of
the international community to resolve disputes peacefully.

We are working with the international community to effect change
in Iran's abysmal human rights record, for example. According to our
own documentation and to international organizations, the government
of Iran uses torture, excessive and lethal police force, and arbitrary de-
tention to repress free speech, freedom of association, and religious
freedom, among other abuses. We are actively seeking a resolution on
the human rights situation in Iran in the U.N. General Assembly's
Third Committee or at the U.N. Commission on Human Rights.

We believe a united international front is especially critical in dealing with Iran's clandestine nuclear weapons program, about which
there is widespread concern across the international community. We
also remain concerned about Iran's biological and chemical weapons
and ballistic missile programs. Our efforts to counter these programs
include bilateral discussions with allies and friends, such as President
Bush's meeting with Russian President [Vladimir] Putin at Camp
David, where the two leaders agreed on the goal of an Iran free of nuclear weapons. We consistently have urged our friends and allies to
condition any improvements in their bilateral or trade relations with
Iran on concrete, sustained, and verifiable changes in Iran's policies
in this and other areas of concern. . . .

# Iran Must End Support for Terrorism

We are also engaged in bilateral and multilateral efforts, from sanctions to direct appeals, to put a stop to Iran's support for terrorist or-

ganizations, which we believe includes [the Islamic terrorist group] al-Qaida. We believe that elements of the Iranian regime have helped al-Qaida and [the radical Kurdish Islamist group] Ansar al-Islam transit and find safehaven in Iran, despite Iran's official condemnation of these groups. Despite public statements that they would cooperate with other countries, the Iranians have refused repeated requests to turn over or share intelligence about all al-Qaida members and leaders they claim to have in custody. As the President made clear, Iran must change its course on this front; resolution of this issue would be an important step in U.S.-Iranian relations and we cannot move forward without this step. We will continue to press this issue from the highest levels of our government, as well as to encourage our friends and allies to press the Iranians.

In its support for terrorism, including by arming violent factions, Iran is interfering in the internal affairs of Afghanistan and Iraq, and especially in the fate of the Palestinian people. Indeed, Iran continues to be the world's foremost state supporter of terrorism, offering financial and logistical support to both Shia and Sunni terrorist organizations, including Hizballah, Hamas, and Palestinian Islamic Jihad. Through these abhorrent groups, Iran destabilizes the region and tries to stymie any movement toward peaceful resolution of the Middle East conflict.

On the other hand, Iran says it wants a stable, unified neighbor in both Afghanistan and Iraq and despite significant unhelpful interference, has taken a few steps in that direction. This includes rhetorical support, by welcoming the end of the oppressive regime of the Taliban [in Afghanistan], which exported drugs, violence, and millions of refugees across the border into Iran. Iran also welcomed the formation of the Iraqi Governing Council [a 25-member provisional authority appointed by occupation forces]. The Iranians have backed up that rhetoric with pledges of material support . . . and they continue to cooperate with regional counter-narcotics and refugee repatriation efforts. . . .

# The Strategy of Encouragement

An important aspect of ongoing U.S. efforts to influence the direction of Iranian policy is encouraging the healthy development of Iran's civil society. We see many signs that the people of Iran want a different life and a more responsive government, and we believe we can encourage such developments through direct engagement with the Iranian public. An estimated 70 percent of the 68 million people in Iran are under the age of 30, and they are far more concerned about Iran's chronic unemployment than they are about Iran's past. Iranian displays of sympathy after the September 11th attacks and polls showing overwhelming desire for improved relations with the U.S. reflect strong popular sentiment, as do demonstrations and elections in support of reform. The government tries to blame any sign of dissent on outside agitators, but it is clear that the agitation in Iran is a genuine expression of a homegrown desire for change. Consider that thousands of ordinary Iranians spontaneously flocked to the airport to greet Shirin Ebadi two weeks ago [in October 2003] when she returned to Tehran after the announcement of the award of the Nobel Peace Prize.

We believe we can encourage the triumph of public resolve by engaging in direct communication with the people of Iran. We are doing this through Radio Farda, which operates 24 hours a day, and Voice of America (VOA) radio and television broadcasts into Iran. VOA has recently instituted a daily Persian television news program to Iran, in addition to its two weekly television feature programs. In May, the State Department brought on line a website in Persian and we continue to explore opportunities to incorporate Iran-related projects into our broader Middle East Partnership Initiative. Our Education and Cultural Affairs Bureau also supports cultural, educational, and professional exchanges. . . .

I firmly believe that our strategy will succeed in helping to push and pull Iran in the right direction, particularly with the close cooperation of other nations. But it is not up to the United States to choose Iran's future. Ultimately, I am most hopeful for that future because it is the people of Iran themselves who are providing the key impetus for change. Despite living under a regime that limits or denies its people even basic human rights, Iranians are engaged in a very rich and lively debate about the kind of society they want for themselves and for their children. They have made it clear that they want democratic and economic reform, accountability and transparency from their government, an end to corruption, religious moderation, and reintegration with the international community. The Iranian people should know of our support for their aspirations, but also that the full rewards of that support will only be realized once their government ends its destructive external and internal policies. We look forward to the day when the will of the people of Iran prevails.

# The United States Has Exaggerated the Threat Posed by Iran

By Michael Donovan

*In the next selection, Michael Donovan identifies and examines the three main accusations leveled at Iran by the United States. They are the following: Iran is pursuing nuclear weapons, it supports international terrorism, and it is interfering in Iraq's reconstruction following its war with the United States. Donovan argues that the threat of Iran's actions has been exaggerated by the Bush administration, whose hostile rhetoric has served to escalate the conflict. The increased tension not only encourages Iran to build up its military capabilities but also creates insecurities on the domestic front that may undermine Iran's moderate reformers. According to Donovan, the United States must end its hostile stance toward Iran and allow reformers to change the country from within. Donovan is a research analyst for the Center of Defense Information, an independent, nongovernmental research organization that monitors the U.S. military. He received a doctorate from the University of Edinburgh, United Kingdom, with a dissertation examining U.S.-Iranian relations and political stability in Iran.*

[In 2003], the rhetoric from Washington about Iran has grown more heated than it has been for many years. According to the officials in the George W. Bush administration, Iran is guilty of a myriad of crimes. The three most prominent accusations focus on Iran's pursuit of nuclear weapons, its support for terrorism, and the probability that it is meddling in the reconstruction of Iraq. All three charges are valid

Michael Donovan, "The Case Against Iran," www.cdi.org, July 12, 2003. Copyright © 2003 by the Center for Defense Information. Reproduced by permission of the author.

to a point. But some of these concerns have been overstated by an administration that is eager to challenge another member of the "axis of evil."[1] Beyond the rhetoric, there is reason to believe that Iran is motivated by an acute sense of insecurity. By stoking these fears, and refusing to take Iran's legitimate security concerns into account, Washington may be pushing Iran in dangerous directions.

# Iran's Nuclear Program

Iran's nuclear program began under the Shah [Reza Pahlavi] in 1974, but was abruptly suspended following the Islamic revolution in 1978–79. The Shah also conducted research into the production of fissile material, but these efforts were suspended during the revolution and the Iran-Iraq war. It was not until 1984 that Ayatollah Khomeini revived Iran's nuclear weapons program. There are some indications that he did so reluctantly, viewing these weapons as amoral. In 1987 and 1988, the reactor sites at Bushehr I and II were damaged by Iraqi air strikes, and progress was again arrested.

Throughout the last decade, U.S. and European analysts have suspected that the Iranian nuclear program may have been reconstituted. With Russian and Chinese help, Iran began to develop a civilian nuclear infrastructure that included a 1,000 megawatt nuclear reactor (not yet complete) at Bushehr, on the Persian Gulf. However, in August 2002, the disclosure of two additional nuclear facilities in Arak and Natanz suggested that Iran's nuclear infrastructure could be configured to produce fissile material for nuclear weapons. The Arak facility was identified as a plant for the production of heavy water and the Natanz facility, once complete, may be for uranium enrichment. Following the public disclosure of the facilities, chief of the International Atomic Energy Agency (IAEA), Mohammed El Baradei, visited the Natanz site at the invitation of Tehran. The IAEA's assessment of the facilities will be published June 19.[2]

Tehran has been careful to follow the letter, perhaps as opposed to the spirit, of the law. Both President Mohammad Khatami and his predecessor, Ali Akbar Hashemi Rafsanjani, have categorically denied that Iran is pursuing nuclear weapons. Iran points out that it is party to the Nuclear Non-Proliferation Treaty (NPT) and is entitled to import nuclear reactors and other technologies under the provisions of the treaty. The IAEA has regularly inspected all of Iran's declared nuclear facilities, reports that it is in full compliance with the NPT, and

1. a term introduced by Bush in his January 2002 State of the Union speech to describe Iran, Iraq, and North Korea's threat to world peace    2. The 2003 report urged Iran to allow continued inspections, but found no violation of treaty agreements.

has found no evidence of a nuclear weapons effort. Following revelations about Iraqi clandestine nuclear facilities in 1991, the IAEA invoked its authority to conduct special inspections of undeclared sites. Hoping to avoid suspicion that it was in violation of the NPT, Iran allowed the IAEA to visit any site upon request. The agency has made several visits to undeclared sites in Iran, but has failed to uncover any nonsanctioned activities. IAEA inspections remain an imperfect mechanism for monitoring clandestine weapons programs, and experts are divided as to the value of these visits. Nevertheless, Iranian officials refer to this inspection record with a mixture of pride and defiance. They point out, with some justification, that other countries in the region with suspected nuclear weapons programs are not so transparent.

The balance of evidence suggests that Iran is determined to develop its civilian nuclear infrastructure to the point where a military applied nuclear program could quickly be put in place. Then, two options are plausible. Iran could attempt to covertly produce the fissile material that would make a weapon possible, or Tehran could legally withdraw from the NPT after providing the requisite 90 days notice, citing "extraordinary circumstances."

What might constitute "extraordinary circumstances"? From Iran's point of view, there is no shortage of threats to potentially justify a nuclear weapons program. In the last three years, the United States has invaded and overthrown two of its neighbors, and officials in Washington are now talking of regime change in Tehran. Though the invasion of Iraq removes one strategic rational for Tehran's interest in weapons of mass destruction, the Iranians still feel obliged to deter Israel, which maintains its own nuclear weapons arsenal and is not a member of the NPT. To the east, the Indo-Pakistani nuclear rivalry continually threatens to spill over Iran's borders.

# Support for Terrorism

Iran has been associated with terrorism since the birth of the Islamic Republic in 1979. However, Iran's reputation as a state sponsor of terror led to increasing degrees of isolation just as its political system was beginning to open. Consequently, a combination of domestic and international pressure led Tehran to repudiate terrorism as a political tactic, with the notable exception of support for what is seen in Iran as the Palestinian resistance to Israeli military occupation.

According to the U.S. Department of State, Iran remained the most active state sponsor of terrorism in 2002. The bulk of accusations about Iran's support for terrorists focus on Tehran's backing of Hezbollah [a Lebanon group of Shi'ite terrorists] and the Palestinian rejectionist groups, Hamas, Palestinian Islamic Jihad, and the Popular Front for the

Liberation of Palestine—General Command. While these groups receive financial and ideological support from Arab populations throughout the region, U.S. intelligence reports assert that elements of the Iranian Ministry of Intelligence and the Revolutionary Guard Corps provide these groups with training, funding, weapons and safe haven.

Like the Arab states, Iran has declared its commitment to the Palestinian cause and often militates against the peace process. Unlike many Arab populations, however, Iranians are not subject to the current of anti-Semitism that pervades the region, though sympathies clearly favor their fellow Muslims in Palestine. Some Iranians are beginning to question, at considerable risk, whether Iran's hard-line stance on the Arab-Israeli question serves the interests of the nation or those of the regime. Indeed, the economic and political costs associated with the strategy are mounting. As they do, the chorus of voices arguing that Iran should limit its support for the Palestinian cause to the political and ideological will also grow. It is for this reason that President Khatami has suggested that Iran would not oppose any settlement that was just and accepted by all parties.

# Iran and al Qaeda

More recently, the Bush administration has accused Iran of harboring and possibly colluding with al Qaeda operatives. U.S. officials are concerned that al Qaeda's security chief, Saif al Adel, possibly had a hand in coordinating the May 2003 terrorist attack in Saudi Arabia while he was hiding in Iran. Other al Qaeda members, as well as members from Ansar al Islam, are thought to have fled over the Iranian border when the United States invaded Afghanistan and Iraq respectively. It is likely that some al Qaeda blended into the millions of Afghan refuges that had fled the Taliban and the fighting. Iran has responded to Washington's accusations by arresting some known al Qaeda operatives and expanding its security net in the eastern provinces.

While U.S. Secretary of Defense Donald Rumsfeld insists that there is "no question but that there have been and are today senior al Qaeda leaders in Iran, and they are busy," unidentified administration officials say there is no concrete intelligence proving that al Qaeda is operating out of Iran. Indeed, a link between Iran and the al Qaeda would be even more surprising than a link between Iraq and the group, which also has yet to be established. Iran is a majority Shia country; al Qaeda operatives subscribe to a fundamentalist interpretation of Sunni Islam. Beyond these doctrinal differences, Iran was a mortal enemy of both Taliban and al Qaeda while the two controlled most of Afghanistan. Al Qaeda operatives played a role in the massacre of Iranian diplomats after the Taliban captured [the Afghan town of] Mazar e Sharif [in August 1999]. Throughout the latter 1990s, Iran backed anti-Taliban groups with substantial military aid.

# Meddling in Iraq

The U.S. government has also asserted that Iran is meddling in the reconstruction of Iraq, but has thus far offered little conclusive evidence to support its case. Nevertheless, it seems likely that Iran is vying for influence among their coreligionists, the Iraqi Shia. According to some analysts, the clerical hardliners fear the example that a democratic Iraq might set. Perhaps more importantly, the Shi'ite seminary in Najaf, once suppressed by [former dictator of Iraq] Saddam Hussein, could begin to challenge its counterparts in the Iranian city Qom. Iraqi Shi'ite clerics generally have not embraced *velayat-e faqih* (rule of jurisprudence), the doctrine from which Iranian clerics derive their legitimacy and right to rule. Consequently, the clerics of Najaf could conceivably call into question the foundation of the Iranian theocracy.

Reports that Iran had used a Shi'ite Iraqi opposition group, the Supreme Council for the Islamic Revolution in Iran (SCIRI), to foment trouble raises additional concerns. Prior to the invasion of Iraq, there was some talk of using this armed group to fight against the regime, but Washington did not trust the group because of its close connections to the Iranian government. Recently, there have been reports that SCIRI insurgents and elements of the Iranian Revolutionary Guard corps have been infiltrating into Iraqi cities to stir up anti-American sentiment. Iranian agents or proxies may also be involved in the tense competition for power and following that is currently unfolding within Iraq's divided, and often hostile, clerical community.

These U.S. concerns are probably valid, but they may also be somewhat overstated. Iranian influence in Iraq is likely to remain somewhat limited, even without the efforts of the United States to exclude it. Iranian and Iraqi Shia may share the same religion, but they are divided by their different ethnicities (Iraqis are Arab, and Iranians are Persians). The fact that the two populations are coreligionists was not enough to keep each from fighting the other throughout the eight-year-long Iran-Iraq war, and nationalism is still an important factor in their respective political cultures. Iraqis may invite Iranian assistance, but they are likely to do everything possible to limit Iranian influence.

As for Iran, few other countries in the region have as much at stake in the future of Iraq. Tehran views the post-conflict reconstruction of both Afghanistan and Iraq as legitimate security concerns, having been threatened by both in the past. Excluding Iran from any role in the future of the two countries has left Tehran with few alternatives beyond acting as a spoiler to U.S. efforts.

With the fall of Saddam Hussein, the United States was offered a historic opportunity to begin integrating Iran into a regional security framework. Unfortunately, Washington has chosen hostility instead. But the Bush administration's policies belie the fact that a nascent, but meaningful, democratic experiment in Iran is yielding a generation of leaders who have less interest in the confrontational policies of the Islamic hardliners. It remains far from clear that bellicose rhetoric and military intervention are the best approaches to the challenges Iran poses. Rather, as General Anthony Zinni, former commander in chief of U.S. Central Command and a Center for Defense Information distinguished military fellow, argues, the key to nonproliferation in Iran is domestic political reform. Iran's support for extremist Palestinian groups and its weapons of mass destruction programs will continue to menace U.S. interests in the region, but tying U.S. policy in the region to an exaggerated perception of the Iranian threat will unsettle allies, push Iran in dangerous directions, and may undermine the moderate political forces that will eventually bring Iran back into the international mainstream.

# Iran-Iraq Relations After the Defeat of Saddam Hussein

By Anoushiravan Ehteshami

*When the United States went to war against Iraq in March 2003, Iran did not offer support to either of its enemies. The U.S. victory three weeks later not only removed Iraq's dictator Saddam Hussein from power but also created a volatile and unstable Iraq. Given Iran's history of conflict with Iraq and continued hostile rhetoric from the United States, Iran faces an uncertain future. The country must not only prepare itself for a possible confrontation with a new Iraqi regime or the United States itself but also keep its options open for possible cooperation with the two countries. In the next article, Anoushiravan Ehteshami contends that although the U.S. war on Iraq has removed Iran's greatest security threat, it has also created a new set of concerns for the Islamic Republic. Ehteshami is a senior lecturer in Middle East politics at the University of Durham in the United Kingdom, and is vice president of the British Society for Middle Eastern Studies. She is the author of* After Khomeini: The Iranian Second Republic *and* Syria and Iran.

Certain myths about Iran's relationship with Iraq must be laid to rest, the first being the persistent notion that the two countries are somehow destined for rivalry. The notion that ancient geopolitical animosities underline relations between the modern states of Iran and Iraq is false. Indeed, when viewing the region through the lens of history rather than contemporary realpolitik, the strategic partnership that has emerged between secular, pan-Arabic Syria and Islamic Iran is difficult to explain. Yet, interest dictates policy, and history informs it— not the other way around. In other words, the tensions in modern Iranian-Iraqi relations have virtually nothing to do with the Ottoman-

Anoushiravan Ehteshami, "Iran-Iraq Relations After Saddam," *Washington Quarterly*, vol. 26, Autumn 2003. Copyright © 2003 by the Center for Strategic and International Studies (CSIS) and the Massachusetts Institute of Technology. Reproduced by permission of the MIT Press, Cambridge, MA.

Persian competition over Mesopotamia or theological and ideological differences between Sunnis and Shi'as, though both sides have routinely ripped pages from the history books to justify their own actions. Unearthing the complexities of contemporary Iranian-Iraqi relations, therefore, requires accepting the fact that tensions between the two countries have their roots in more recent developments.

# The Iranian Revolution

Among the factors that have overwhelmingly influenced Iranian-Iraqi relations, the 1979 Iranian revolution is one of the most important. It is not surprising that relations would be tense between a revolutionary, clerical, Shi'a-dominated Iran and an Arab, nationalist-secular, Sunni-dominated, one-party dictatorship ruling over Iran's only Shi'a-majority neighbor in Iraq. Even though the revolution removed Baghdad's strongest Gulf rival and one of the West's strongest regional allies, Muhammad Reza Shah Pahlavi, from Iran, Iraq's response was understandably less than sanguine, not sharing in Iran's jubilation.

The new revolutionary leadership in Tehran inherently challenged the new Iraqi president, Saddam Hussein, who had only taken control of the Iraqi regime months before in July 1979. Iraq was also forced to trade the known quantity of the shah for the unpredictability of Ayatollah Ruhollah Khomeini. Ironically, Iranian fears of Iraqi hostility against the Islamic regime, informed by Khomeini's assessment of the Iraqi regime that had acted as his host in the holy city of Najaf from 1965 to 1978, mirrored Iraqi mistrust of Iran. Through an accident of history, personal mistrust inflamed political tensions, as Khomeini had experienced firsthand the systematic suppression of the Shi'a clerical establishment and its flock by the Ba'th [Iraqi] leadership in the 1960s and 1970s. In Khomeini's eyes, Saddam himself had been implicated in the regime's anti-Shi'a campaign even before rising to the pinnacle of power in Iraq.

Since its inception in 1979, Iraq has been one of the Islamic Republic's main foreign policy challenges. Iraq not only challenged Iran's regional ambitions before the revolution but also did its best to isolate it from the Arab world, posing a direct security threat to its territory, economy, and population. Iraq had engaged in the destabilization of the border region soon after the victory of the revolution, for example, and had started shelling Iran's strategic economic targets well before its invasion of Iranian territory in September 1980.

Despite the tensions in Iraqi-Iranian relations during the shah's reign, the revolution in Iran did not ease the atmosphere between the two countries. With the uncompromising Khomeini in charge, the new and inexperienced Iranian leadership, also drenched in revolutionary fervor, found it almost impossible to resist the temptation to taunt Saddam and challenge his regime's legitimacy in a pointed and public manner.

## The Iran-Iraq War

The eight years of fighting that followed the revolution and the geopolitical changes that occurred in its wake merely underlined the depth of animosity between the revolutionary regime in Iran and Saddam's government. Personality clashes, geopolitical rivalries, regime types, and deep suspicion at the leadership level combined to escalate a manageable border dispute into a more general conflict, which resulted in all-encompassing interstate war. The war . . . was sustained by the hegemonic instincts of Tehran and Baghdad, both of which desired to be the dominant Gulf power. The war was ultimately about territory, influence, and survival—it was not about religion or some historically rooted difference.

Iran ultimately lost the war for two primary reasons: its blunders on the battlefield as well as in the diplomatic arena and the strategic as well as political support that the United States and its allies were prepared to lend Iraq in its campaign against the Islamic Republic. Iran at that time was considered to be an irredentist power bent on redrawing the strategic map of the region in its own revolutionary Islamic image. As a result, Saddam managed to leverage that threat and outlast both of his twentieth-century foes, Khomeini and U.S. president George H.W. Bush, by skillfully turning Iraq's geopolitical weaknesses into military virtues. Iraq used its maritime handicap and vulnerabilities, for instance, to secure the use of France's antiship Exocet missile system on the Super-Etendard platform for attacking Iranian shipping the length and breadth of the Gulf. In a period of four years from 1984, it systematically attacked commercial shipping and military targets and forced Iran to respond by attacking neutral or Iraq-bound shipping traffic. Baghdad's military responses to its geopolitical vulnerabilities, in short, had given birth to the "Tanker War."

The war between these two major oil producers created a host of policy dilemmas for the energy-hungry United States, which had just lost its most reliable regional partner to Islamist revolutionaries in Iran and was concerned that the revolutionary storm from Iran might shake the foundations of [Saudi Arabia's rulers] the House of Saud (the other important U.S. ally) as well as the smaller and more vulnerable Gulf Arab states. Although the war would check the power of Iran's revolutionaries, Washington did not cherish the prospect of the Iran-Iraq War spinning out of control and affecting the stability of the entire region. That would be too high a price to pay for the containment of the Iranian revolutionaries. Nonetheless, in the absence of one of its main security twin pillars in the Gulf, the United States had little option but to increase its military commitment to the region while taking advantage of the war to check Iran's ambitions and expand U.S. influence in Baghdad.

U.S. fear of the Iranian revolution caused a change in U.S. policy toward the war in 1980. Essentially, it used the war and the wider security crisis in the Gulf as an opportunity to extend its reach and consolidate its partnerships with several Gulf Arab states. Washington tightened its sanctions regime on Iran while slowly shifting its weight behind Baghdad. Iraq's Gulf Arab backers feared the growing influence of the Iranian revolution and saw Iraq as the first line of defense against revolutionary Iran, further encouraging the U.S. shift toward Iraq and the other Arab states. The only U.S. support for Iran during the Iran-Iraq War came during the Iran-contra affair of 1986–1987, when the United States covertly supplied Iran with badly needed war material (HAWK missiles, TOW antitank missiles, and spare weapons parts) in exchange for the freedom of U.S. citizens held hostage in Lebanon. Despite some significant policy differences with Baghdad and Iraq's "accidental" missile attack on the USS *Stark* in May 1987, Washington maintained its pro-Iraq stance until the end of the war in 1988.

# The Lasting Legacy

The absence of a formal peace treaty with Iraq since the end of hostilities in 1988 has intensified Iran's policy challenges toward Baghdad. More recently, Iraq's invasion of Kuwait in 1990 reinforced Tehran's perceptions of Saddam's Iraq as a politically challenging and, possibly, militarily superior neighbor. Along with the United States, Iran was unable to find suitable solutions to the range of security challenges presented by Iraq, including its possession of weapons of mass destruction (WMD), territorial encroachment on neighboring states and Iran itself, potential to disintegrate into a vacuum of ministates with huge geopolitical consequences, and powerful position in the oil market.

Yet, Iran had to face other challenges unilaterally, such as Baghdad's support for armed Iranian opponents of Iran's government, the sociopolitical effects of hosting a large Iraqi exile community, and the influence of domestic Iraq ethnic divisions on Iran. The depth of Iran's problems with Iraq stands in stark contrast to Tehran's singular failure to deliver a consistent set of policy options toward Baghdad. Instead, it has taken a shortsighted approach toward Iraq.

Many Iranians now in positions of power and influence served on the front lines and still speak bitterly of the war years. They openly curse Saddam for the damage inflicted on their country and for the misery he brought them, their families, and associates. Iranians generally regard the war as the root cause of the economic and social problems their country faces today; by extension, they have held the former Iraqi regime responsible for these difficulties for the last decade. To some extent, this idea of Iraq as the source of all evil is a fig leaf that disguises incompetence and corruption at home; capable technocrats argue reasonably, however, that the socioeconomic legacy of the 1980s has undermined the attempts of Iranian president Ali Akbar Hashemi Rafsanjani (1989–1997) and Muhammad Khatami (1997–present) to introduce socioeconomic and political reforms. . . .

The debate about the conduct of the war further contributes to popular Iranian perspectives on, as well as Iran's more formal policy toward, Iraq. A growing body of opinion in Iran holds that Tehran's unconditional acceptance of United Nations Security Council Resolution 598, which produced a cease-fire in July 1988, marked only a dubious victory for the Islamic Republic because, in the end, none of Iran's war aims were realized. Iran had failed to topple the Iraqi regime, to secure a border treaty with Iraq, or to extract war reparations in exchange for a cease-fire.

Such revisionist views of the Iran-Iraq War form a new battleground for the main Iranian power blocs. As "public ownership" of policy toward Iraq has increased—where motion pictures, documentaries, and even war memories fuel the debate about the war—Iranians are openly articulating their views on the best course for their country in the aftermath of the U.S. military assault of Iraq: whether U.S. action will serve Iran's broader long-term interests or whether Iran should support the return of UN personnel to Iraq and oppose unilateral U.S. action. Notably, 127 members of parliament out of 286 legislators in Iran penned an open letter to Supreme Leader Ayatollah Ali Hoseini Khamenei in May 2003 asking him to "drink from the chalice of poison" and allow for the broadening of the reform process at home as well as a comprehensive review of Iran's relations with the United States. The debate is raging now and will have direct policy consequences.

For Iran, another social legacy of the war has been a culture of remembrance and commemoration including the fountain of blood (oozing red liquid) at the entrance to Tehran's main war cemetery, streets named after war heroes, and the regularity with which key dates and events of the war are marked. Although state-level exchanges between Tehran and Baghdad had become commonplace as early as the mid-1990s, and the physical scars of eight years of conflict had all but disappeared, state and civil society structures in Iran have combined to perpetuate more subtle reminders of the war and anger toward Iraq for its past misdeeds. As a result, until the fall of Saddam's regime, no government spokesperson could openly support reassessing Iranian attitudes toward Iraq. The presence of some 600,000 Iraqi refugees of various ethnic and political backgrounds in Iran has also kept Iranians' interest in, and awareness of, developments across the border very much alive.

In the security realm, Iran's defense strategies and many of its military purchases have continued to reflect a preoccupation with Iraq as a potential enemy. At both the theoretical and practical levels, Iran had been preparing, if only subconsciously, for another encounter with Iraq. Iran's military acquisition and development programs have been dominated by perceived shortcomings exposed by the war during the 1980s, including weaknesses at sea (as revealed by the "tanker war" and encounters with the U.S. Navy in 1987), in air defense (manifested in the "war of the cities" and in Iraq's superior airport and ability to strike at strategic targets), in the maneuverability of ground forces, and in deterrence. Tehran remains extremely concerned about Iraq's WMD potential, which of course guided Iranian defense purchases until recently, and has developed sophisticated surface-to-surface missiles and imported Russian-supplied long-range strike aircraft, at least in part as a counterforce to those unconventional Iraqi weapons. It has also been building up air defense systems around strategic targets. In the wake of Saddam's regime, however, Iran's military machine is, on balance, far larger and more sophisticated than Iraq's. Tehran failed to anticipate the virtual destruction of the Iraqi war machine arsenal, as well as Iraq's placement under the protection of a country with a far superior military—the United States.

Just as the strategic relationship between Damascus and Tehran since 1980 has arisen out of shared objectives and fears, the distance between Tehran and Baghdad has not been bridged because of mutual suspicion and fear. Yet, events show that this suspicion and fear is relatively young—not ancient—and rooted in actual, everyday quality-of-life issues and practical, strategic interests, not a deeply embedded ideological opposition.

Moreover, although the geopolitical realities which continue to divide Iran and Iraq should not be underestimated, the two neighbors have demonstrated a remarkable capacity, despite lasting tensions, for bilateral cooperation in pursuit of each of their interests in Gulf security since the end of their war in 1988. The two countries reestablished diplomatic relations, rebuilt some of their old economic ties, and broadened intergovernmental exchanges during the 1990s on the issues of war reparations, their common border, and prisoners of war. If Iran could deal with a more moderate political leadership in Iraq instead of Saddam, they should be able to develop this relationship further.

# A Change in U.S. Policy

Although the UN declared in late 1990 that Iraq was the aggressor in the Iran-Iraq War, setting the stage for a cold peace that would hang over Iran and Iraq for most of the decade, Iraq's invasion of Kuwait had so weakened the position of Iraq as a regional actor and the Ba'thist regime internally that Tehran thought it impractical to seize on the UN's declaration to push for reparations and an Iraqi admission of guilt for starting the war. Instead, Tehran chose to isolate the pariah regime in Iraq diplomatically, maintaining concern that dealing with Saddam would only help him domestically and hoping that Saddam would be overthrown. Thus, Iran chose not to realize one of its key war aims, a return to the 1975 Algiers accords (which had delineated the border between Iran and Iraq and had provided the basis for co-operative relations between them), even though Saddam was now offering it as the basis for negotiating a new border agreement with Iran.

The 1990s marked continuing steady tensions between Iran and Iraq despite the emergence of a U.S. policy that should have brought the two nations closer together. The Clinton administration marked the onset of a policy of dual containment toward Iran and Iraq as part of the new U.S. designation of rogue states. The United States, seemingly tired of playing the two countries against each other and having been burned by both, adopted a policy of dual isolation, in which neither would be assisted and both would be pressured to conform to international norms considered vital for the preservation of regional and international security. Although Iraq was technically the only one of the two under U.S. and UN sanctions, dual containment effectively placed Iran and Iraq in the same boat, despite the many significant differences between the two countries' political systems and socioeconomic compositions. Throughout the 1990s, the two neighbors continued to view each other and not the United States as their greatest source of insecurity. The Iranian armed forces remained fearful of the Iraqi regime's posturing toward Iran and had contingency plans for renewed Iraqi provocations over the Shatt al Arab border issue.

With the departure of the Clinton administration in 2000 and the erosion of dual containment as the European powers, Russia, and China deepened their diplomatic and trade links with Iran, Tehran hoped that a better working relationship could be established with the new Republican White House. It was an open secret in Tehran that the leadership expected better relations with the Republicans in particular. Despite some evidence of flexibility on both sides (during the campaign, Governor George W. Bush's team focused more on Iraq as a foreign policy problem than Iran, for example, and Tehran let it be known that it hoped a president from the United States' own oil state would better understand the complexities of the Gulf region), Iran's anxiety was heightened in 2002 when it found itself portrayed by the new U.S. president as Iraq's bedfellow—this time, in an "axis of evil." Only this time, whereas dual containment had sought to isolate Iran and curtail its regional influence, Tehran calculated that the new doctrine targeted specific ruling regimes as "evil" powers and potentially subjected the Iranian leadership to direct U.S. pressure.

When the United States eventually implemented its new doctrine and took military action against Iraq, Iran, which under other circumstances would have welcomed any effort to remove the Iraqi regime, was unprepared and unwilling to lend any direct support to the U.S. effort. The reason was simple and understandable: Iran itself was in the U.S. crosshairs as an evil power. It was no longer sufficient for Iran to be in the containment zone (à la the Clinton doctrine), and the Bush administration introduced the axis of evil concept that proved to be a far more aggressive doctrine. Why should Iran help overthrow Saddam when rapid success may have facilitated U.S. efforts then to overturn the regime in Tehran? . . .

Saddam's fall will affect factional rivalries in Iran. Some elements in Iran will point to U.S. behavior in Iraq—the apparent renewed support for the Iraq-based anti-Tehran Mujahideen-e Khalq organization, the imposition of a U.S. political model on a Muslim state, the establishment of military bases, and the control of Iraq's oil wealth—as well as the expansion of military facilities in the small Gulf Arab states of Bahrain and Qatar and the perceived encirclement of Iran through an elaborate network of alliances—as justification to encourage some Iraqi Shi'a forces to assist Tehran in extending its power in Iraq by infiltrating the emerging post-Ba'thist polity. Tehran does have a potentially powerful ally among Iraqi Shi'as . . . who regularly mounted military and logistical operations in Iraq during Saddam's rule.

Tehran has also been heavily engaged in training and maintaining the [anti-Saddam, Shi'a] al-Hakim group as well as the well-established Kurdish Patriotic Union of Kurdistan [PUK] and the Islamist [anti-American] al-Da'wa party. As the [Iran-based Supreme Council for the Islamic Revolution in Iraq] SCIRI gets embedded in Iraq itself, however, Tehran's grip over it is bound to loosen, particularly because SCIRI's leadership will have to strike compromises with an emerging Iraqi leadership if it is to remain a force in the post-Saddam power structure. Another possibility is that Iranian control of SCIRI could bring Iraqi Shi'a influences into Iran and encourage fresh thinking on Shi'a issues, thereby endangering the semi-unity of the religious establishment in Iran over matters of state (such as the future role of the Faqih [Islamic judiciary], the clergy's future role in day-to-day affairs, curtailment of the Faqih's constitutional powers, and relations with the United States) and national political issues (such as the distribution of power between the three branches, social and political reforms, freedom of the press, and organization of political parties).

Those in Tehran who are deeply worried about developments in Iraq and the domestic and foreign policy consequences of manipulating Iraq's large Shi'a community for narrow political ends counsel caution. Far from seeking to meddle in Iraq's internal affairs, they desire to protect [the Iraqi seminary town of] Qom's place as the beating heart of Shi'ism.

They also wish to use the opportunity afforded by Saddam's overthrow to deepen relations with the Gulf Cooperation Council [GCC] countries (Bahrain, Kuwait, Oman, Qatar, the United Arab Emirates, and Saudi Arabia). The end of Saddam's regime has removed a stiff barrier to closer Iranian links with the GCC states. Tehran no longer has to worry about the GCC states keeping their distance in fear of Iraqi pressure, and the fall of Baghdad has allowed for the emergence of the Shi'a issue into the open. The fear that Saddam's removal would somehow lead to the rise of an Iranian-controlled, Shi'a-dominated state in Iraq, as expressed in 1991, has not come to pass, and the Shi'a dimension of Iraqi society is no longer seen as a security threat but rather a part of the country's reality. The Shi'as no longer stand in the way of closer relations between Tehran and the GCC states. U.S. removal of the Ba'thist regime in Iraq has allowed Arab Shi'ites in that country to make their presence known, and Iran no longer has to fear negative fallout in the Arab world from its own association with this community.

# The Debate in Iran

Iran's moderates and pragmatists point to the rapid dismantling of U.S. military deployments in Saudi Arabia as proof that Washington has no intention of targeting Iran and further argue that the United States may well be ready for inclusive discussions about collective security arrangements in this vital subregion. Tehran, they argue, should maintain its steady course of détente and take advantage of the new situation to underline its cooperative nature and enter into deeper dialogue with the United States as well as the EU about the future shape of the Gulf security framework. They see an extended role for Iran in helping to reduce sources of tension in the Gulf as in their national interest.

As already noted, however, the liberation of Iraq and of the Shi'a communities within it could widen Iran's own political fault lines. For this reason, Iran's leaders will struggle to balance the adventurous tendencies in Iran that desire to take advantage of the confusion in Iraq, to penetrate and control its Shi'a establishment, against the deeply conservative and cautious instincts of the majority who wish to avoid danger by adopting a minimalist posture. Although Iran will find it impossible to distance itself entirely from the Iraqi Shi'as and is likely to try to exert its influence in post-Saddam Iraq, it does recognize that it can only pursue its aims within a rapidly changing regional geopolitical environment and strategic setting.

# Facing New Regional Realities

Effectively, Tehran and others in the region must accept that unchallenged U.S. force has removed the greatest source of insecurity to the Gulf (and to Iran in particular), and in doing so, the regional balance of power has again been shifted as a consequence of U.S. action. They must also recognize opportunities for greater investment and commerce across the Middle East that are emerging from Washington's operation in Iraq and its carefully laid-out road map for the resolution of the Palestinian/Arab-Israeli conflict.

In this new environment, Iran faces a stark choice: either continue to resist U.S. penetration of the region by heavy investment in what has become a shrinking circle of allies or exploit its considerable tactical advantages to broaden its policy of détente and diplomacy for greater economic and political gains. Washington's behavior and its decision on what it means for Iran to be one of only two remaining members of an axis of evil will of course partly determine which path Tehran can choose. Iranian concerns about being pressured to accept the road map in the Arab-Israeli conflict or about Iraq's next government being a U.S. puppet regime may manifest themselves in Iranian foreign policy initiatives with such rejectionist actors as Syria, Hizballah [a militant, Lebanon-based Islamic group], and perhaps other willing partners feeling the chill from the U.S. presence in the new regional order.

More broadly, when it comes to relations with the "Great Satan,"[1] ideology more than policy tends to define the place of the United States in Iran's agenda. Tehran still clearly separates its bilateral concerns with the United States from any potential common interests in Iraq, a separation that is favorable to the U.S.-Iraq strategy but does not assuage Iranian perceptions of the United States as the leader of a cultural invasion. Thus, although the United States has finally delivered on the most important of Iran's goals in its eight-year war with Iraq—the removal of the Ba'thist regime—the tensions between Washington and Tehran have presented the removal of the Iraqi regime as a new "poisoned chalice" with which Tehran must contend.

---

1. a derogatory term used by some Muslims to refer to the United States

# 🔥 CHRONOLOGY

## 1906
The shah signs a constitution that limits royal power and establishes an elected parliament.

## 1921
Military commander Reza Khan seizes power.

## 1926
Reza Khan becomes king and adopts the title Reza Shah Pahlavi.

## 1935
Formerly known by its English name, Persia, "Iran" is adopted as the country's official name.

## 1941
British and Russian forces occupy Iran; Reza Shah Pahlavi is deposed in favor of his son, Shah Mohammad Reza Pahlavi.

## 1951
**March:** Parliament votes to nationalize the oil industry.
**April:** Mohammad Mosaddeq is appointed prime minister; the United Kingdom boycotts the purchase of Iranian oil.

## 1953
Mosaddeq is ousted; the shah resumes power.

## 1963
The "White Revolution" is launched by the shah.

## 1964
Ayatollah Ruhollah Khomeini is exiled.

## 1978
The shah leaves Iran following widespread demonstrations and strikes.

## 1979
**February:** Khomeini returns; Mehdi Bazargan becomes prime minister of the provisional government.

**April:** The Islamic Republic of Iran is established.
**November:** The U.S. embassy in Tehran is seized.
**December:** Referendum approves new constitution.

## 1980

**January:** Abolhasan Bani-Sadr is elected the first president of the Islamic Republic.
**April:** A U.S. rescue mission ends in disaster in a sandstorm in the Iranian desert.
**July:** The exiled shah dies of cancer in Egypt.
**September:** Iraq invades Iran.

## 1981

**January:** American hostages are released after 444 days in captivity.
**October:** Ali Khamenei is elected president.

## 1984

**January:** The U.S. State Department adds Iran to a list of nations supporting terrorism.
**February:** The "Tanker War" begins, affecting all shipping in the Persian Gulf; evidence surfaces of Iraqi use of chemical weapons against Iran.

## 1986

The United States begins to covertly ship arms to Iran in exchange for the release of U.S. hostages held in Lebanon.

## 1988

The U.S. cruiser *Vincennes* shoots down an Iran Air flight over the Persian Gulf, killing all 290 people on board; Iran accepts a cease-fire agreement with Iraq.

## 1989

**February:** Khomeini issues a religious edict (fatwa) ordering Muslims to kill British author Salman Rushdie for his novel *The Satanic Verses*, considered blasphemous to Islam.
**June:** Ayatollah Khomeini dies; President Khamenei is appointed Supreme Leader.
**September:** Iran and Iraq resume diplomatic relations.

## 1995

**March:** Russia signs a deal to provide Iran with a nuclear reactor.
**April:** Washington announces a ban on all direct U.S. trade with Iran.

# 1997

Mohammad Khatami is elected president of Iran with 70 percent of the vote.

# 1998

**January:** President Khatami calls for a "dialogue with the American people."

**September:** Iran deploys thousands of troops on its border with Afghanistan after the Taliban regime admits killing eight Iranian diplomats and a journalist in Mazar-e Sharif; Iran officially drops the fatwa against Rushdie.

# 1999

**May:** The United Kingdom and Iran restore full diplomatic ties.

**July:** Demonstrations at Tehran University over the closure of reformist newspaper *Salam* leads to six days of rioting and the arrest of over a thousand students.

# 2000

Reformists win two-thirds of the seats in parliamentary elections.

# 2001

President Mohammad Khatami is reelected with 77 percent of the vote.

# 2002

In his State of the Union address, U.S. president George W. Bush introduces the term "axis of evil" to describe Iran, Iraq, and North Korea as three nations that threaten world peace.

# 2003

**June:** Thousands attend student-led protests in Tehran against clerical establishment.

**October:** Shirin Ebadi becomes Iran's first Nobel Peace Prize winner.

**November:** UN nuclear watchdog, the International Atomic Energy Agency, concludes that there is no evidence of a nuclear weapons program in Iran.

**December:** Forty thousand people are killed and the city of Bam devastated in an earthquake.

# 2004

Conservatives gain control of parliament in controversial elections after the Council of Guardians disqualifies thousands of reformist candidates.

# 🔥 FOR FURTHER RESEARCH

## Books

Ervand Abrahamian, *Iran Between Two Revolutions*. Princeton, NJ: Princeton University Press, 1982.

———, *Khomeinism: Essays on the Islamic Republic*. Berkeley and Los Angeles: University of California Press, 1993.

Fariba Adelkhah, *Being Modern in Iran*. New York: Columbia University Press, 2000.

K.L. Afrasiabi, *After Khomeini: New Directions in Iran's Foreign Policy*. Boulder, CO: Westview, 1994.

Hamid Algar, trans., *Constitution of the Islamic Republic of Iran*. Berkeley, CA: Mizan, 1980.

Ali M. Ansari, *Iran, Islam, and Democracy: The Politics of Managing Change*. London: Royal Institute of International Affairs, 2000.

Hossein Bashiriyeh, *State and Revolution in Iran, 1962–1982*. New York: St. Martin's, 1984.

Asef Bayat, *Street Politics: Poor People's Movements in Iran*. New York: Columbia University Press, 1998.

Hamid Dabashi, *Close Up: Iranian Cinema, Past, Present, and Future*. New York: Verso, 2001.

Mark Downes, *Iran's Unresolved Revolution*. Burlington, VT: Ashgate, 2002.

John L. Esposito and R.K. Ramazani, eds., *Iran at the Crossroads*. New York: Palgrave, 2001.

Henner Furtig, *Iran's Rivalry with Saudi Arabia Between the Gulf Wars*. Reading, UK: Ithaca, 2002.

Shireen T. Hunter, *Iran After Khomeini*. New York: Praeger, 1992.

Ramin Jahanbegloo, ed., *Iran: Between Tradition and Modernity*. Lanham, MD: Lexington, 2004.

Efraim Karsh, *The Iran-Iraq War, 1980–1988*. Oxford, UK: Osprey, 2002.

Homa Katouzian, *Iranian History and Politics: The Dialectic of State and Society.* New York: RoutledgeCurzon, 2003.

Joseph A. Kechichian, *Iran, Iraq, and the Arab Gulf States.* New York: Palgrave, 2001.

Nikki R. Keddie, *Modern Iran: Roots and Results of Revolution.* New Haven, CT: Yale University Press, 2003.

Nikki R. Keddie and Rudi Matthee, eds., *Iran and the Surrounding World: Interactions in Culture and Cultural Politics.* Seattle: University of Washington Press, 2002.

Stephen Kinzer, *All the Shah's Men: An American Coup and the Roots of Middle East Terror.* Hoboken, NJ: John Wiley, 2003.

Christin Marschall, *Iran's Persian Gulf Policy: From Khomeini to Khatami.* New York: RoutledgeCurzon, 2003.

Rudi Matthee and Beth Baron, eds., *Iran and Beyond: Essays in Middle Eastern History in Honor of Nikki R. Keddie.* Costa Mesa, CA: Mazda, 2000.

Mohsen M. Milani, *The Making of Iran's Islamic Revolution: From Monarchy to Islamic Republic.* 2nd ed. Boulder, CO: Westview, 1994.

Ali Mohammadi, ed., *Iran Encountering Globalization: Problems and Prospects.* New York: RoutledgeCurzon, 2003.

Ali Mohammadi and Anoushiravan Ehteshami, *Iran and Eurasia.* Reading, UK: Ithaca, 2000.

Faisal bin Salman al-Saud, *Iran, Saudi Arabia and the Gulf: Power Politics in Transition, 1968–1971.* New York: I.B. Tauris, 2003.

James Francis Warren, *Iranun and Balangingi: Globalization, Maritime Raiding, and the Birth of Ethnicity.* Singapore: Singapore University Press, 2002.

Robin Wright, *The Last Great Revolution: Turmoil and Transformation in Iran.* New York: Knopf, 2000.

Dariush Zahedi, *The Iranian Revolution Then and Now: Indicators of Regime Instability.* Boulder, CO: Westview Press, 2000.

## Periodicals

Mahan Abedin, "Iran After the Elections," *Middle East Intelligence Bulletin*, February/March 2004.

Ali Abootalebi, "The Struggle for Democracy in the Islamic Republic of Iran," *Middle East Review of International Affairs*, September 2000.

Peter Ackerman and Jack DuVall, "The Nonviolent Script for Iran," *Christian Science Monitor*, July 22, 2003.

David Albright, "An Iranian Bomb?" *Bulletin of the Atomic Scientists*, July/August 1995.

Tariq Ali, "Operation Iranian Freedom," *Nation*, July 31, 2003.

Jahangir Amuzegar, "Trouble in Tehran," *Foreign Affairs*, January 27, 2004.

Darius Bazargan, "Iran: Politics, the Military, and Gulf Security," *Middle East Review of International Affairs*, September 1997.

Massimo Calabresi, "Iran's Nuke Admission: What the Country Is—and Isn't—Telling About Its Nuclear Program," *Time*, November 17, 2003.

Patrick Clawson, "The Paradox of Anti-Americanism in Iran," *Middle East Review of International Affairs*, March 2004.

Robert Dreyfuss and Laura Rozen, "Still Dreaming of Tehran," *Nation*, March 25, 2004.

Michael Eisenstadt, "Russian Arms and Technology Transfers to Iran: Challenges for the United States," *Arms Control Today*, March 2001.

Seymour M. Hersh, "The Iran Game: How Will Tehran's Nuclear Ambitions Affect Our Budding Partnership?" *New Yorker*, December 3, 2001.

Eric Hooglund, "Khatami's Iran," *Current History*, February 1999.

Nikki R. Keddie, "Iran: Understanding the Enigma: A Historian's View," *Middle East Review of International Affairs*, September 1998.

Joe Klein and Adam Zagorin, "The Iran Connection," *Time*, July 26, 2004.

Scott Macleod, "All Out of Reforms: A Crisis in Tehran," *Time*, February 16, 2004.

Daniel Neep, "Iran: Changing the Paradigm," *Royal United Services Institute for Defence Studies*, October 2002.

Neil Patrick, "Iran and the United States: The Waiting Game," *Royal United Services Institute for Defence Studies*, July 2001.

Michael Rubin, "More Floggings and Inflation: The Fruits of Reform in Iran," *Daily Telegraph*, April 9, 2002.

Philip Shenon, "President Says U.S. to Examine Iran-Qaeda Tie," *New York Times*, July 20, 2004.

Adam Tarock, "The Struggle for Reform in Iran," *New Political Science*, September 1, 2002.

## Web Sites

*Iran Bulletin: Middle East Forum*, www.iran-bulletin.org. This Web site is the online version of the *Iran Bulletin* journal. It includes articles covering Iran from a democratic, secular, and socialist perspective.

Iran Chamber Society, www.iranchamber.com. A comprehensive site covering everything Iranian, from early history to contemporary cinema, traditional recipes to political leaders.

*Iran Daily*, www.iran-daily.com. The Web site of the reformist English-language Iranian daily newspaper.

Middle East Media Research Institute: Iran, www.memri.org/iran.html. An American Web site that provides analysis of Iranian news items, with a focus on international relations.

Reza Pahlavi's Secretariat, www.rezapahlavi.org. The Web site of Reza Pahlavi, the son of Mohammed Reza Pahlavi, the late shah of Iran. It includes an archive of Pahlavi's speeches and interviews and regularly updated links to news stories about Iran.

Student Movement Coordination Committee for Democracy in Iran, www.daneshjoo.org. This site offers news, press releases, and information about current campaigns and protests. It also includes a good selection of links.

*Tehran Times*, www.tehrantimes.com. The Web site of the conservative English-language Iranian daily newspaper.

World Factbook: Iran, www.cia.gov/cia/publications/factbook/geos/ir.htm. This site is published by the Central Intelligence Agency and provides extensive geographic and demographic information as well as an overview of the structure of Iran's political institutions and economy.

# 🔥 INDEX

United States
    debate in Iran over dialogue with,
        114
    vs. European Union, in views of
        Iran, 79
    has exaggerated threat posed by
        Iran, 100–105
    must continue dual policy toward
        Iran, 95–99
    nuclear arsenal of, as threat to Iran,
        84
    relations with Iran
        after fall of Hussein, 115
        after Iranian Revolution, 8,
            14–15
        after World War II, 8
    role of, in Iran-Iraq War, 12, 37–38
USS *Samuel B. Roberts,* 38
USS *Stark,* 38, 109
USS *Vincennes,* 39

Vance, Cyrus, 25
Vaziri, Haleh, 61
Vilayati, Ali Akbar, 39

Voice of America (VOA), 99

Waite, Terry, 88
Weir, Benjamin, 88
Western civilization
    Iran can benefit from, 73
    Iran's struggle with, 70–72
"White Revolution" (1961), 10
Wolfsthal, Jon B., 82
women
    defiance of clerical establishment
        by, 77
    struggle for rights by, 55–60
Woolsey, James, 87

Yazdi, Ayatollah Mesbah, 63
Yazdi, Ebrahim, 46
youth, role of, in democratic
    reforms, 61–68, 76–77
Yussefi-Eshkevari, Hassan, 64

Zahedi, Ardeshir, 25
*Zanan* (magazine), 56
Zinni, Anthony, 105